SPACE-TIME PROCESSING FOR CDMA MOBILE COMMUNICATIONS

THE KLUWER INTERNATIONAL SERIES
IN ENGINEERING AND COMPUTER SCIENCE

SPACE-TIME PROCESSING FOR CDMA MOBILE COMMUNICATIONS

PIETER VAN ROOYEN
Alcatel Research Unit for Wireless Access
University of Pretoria, South Africa

MICHIEL LÖTTER
Allied Technologies Limited, South Africa

DANIE VAN WYK
CSIR Defencetek, South Africa

Kluwer Academic Publishers
Boston/Dordrecht/London

Distributors for North, Central and South America:
Kluwer Academic Publishers
101 Philip Drive
Assinippi Park
Norwell, Massachusetts 02061 USA
Telephone (781) 871-6600
Fax (781) 871-6528
E-Mail <kluwer@wkap.com>

Distributors for all other countries:
Kluwer Academic Publishers Group
Distribution Centre
Post Office Box 322
3300 AH Dordrecht, THE NETHERLANDS
Telephone 31 78 6392 392
Fax 31 78 6546 474
E-Mail services@wkap.nl>

 Electronic Services <http://www.wkap.nl>

Library of Congress Cataloging-in-Publication

van Rooyen, Pieter.
 Space-time processing for CDMA mobile communications / Pieter van Rooyen,
Michiel Lötter, Danie van Wyk.
  p. cm. -- (The Kluwer international series in engineering and computer science ; SECS 544)
 Includes bibliographical references and index.
 ISBN 0-7923-7759-1
 1. Code division multiple access. 2. Mobile communication systems. 3. Adaptive
antennas. I. Lötter, M. N. II. van Wyk, Danie. III. Title. IV. Series.

TK5103.45. V36 2000
621.3845--dc21

00-020321

Copyright © 2000 by Kluwer Academic Publishers.

All rights reserved. No part of this publication may be reproduced, stored in a retrieval system or transmitted in any form or by any means, mechanical, photo-copying, recording, or otherwise, without the prior written permission of the publisher, Kluwer Academic Publishers, 101 Philip Drive, Assinippi Park, Norwell, Massachusetts 02061

Printed on acid-free paper.

Printed in the United States of America

Contents

List of Figures	xi
List of Tables	xvii
Preface	xix

1. INTRODUCTION   1
   - 1.1 IMT-2000   2
   - 1.2 IMT-2000 Spectrum Allocation   4
   - 1.3 From Second Generation to Third Generation   5
   - 1.4 Universal Mobile Telecommunications System   7
   - 1.5 Third Generation and Beyond   9
     - 1.5.1 Re-configurable Terminals   9
     - 1.5.2 Application and Service Download   11
     - 1.5.3 Broadband Satellite Systems   11
     - 1.5.4 Space-time Techniques   11
   - 1.6 Space-time Processing Systems   12
   - 1.7 Organization of Book   13

2. BASIC SPACE-TIME ASPECTS   17
   - 2.1 Relevant Cellular Terminology   18
   - 2.2 Basic Adaptive Array Concepts   23
   - 2.3 Space-time Processing   28
     - 2.3.1 Space-time Base Stations   34
     - 2.3.2 Space-time Mobiles   35
     - 2.3.3 Advantages of Space-time Processing   35
   - 2.4 System Model for Space-time Processing   37
   - 2.5 Practical Implementation of CDMA Smart Antenna Systems   40
   - 2.6 Summary   41

3. ASPECTS INFLUENCING SPACE-TIME PERFORMANCE   43
   - 3.1 Propagation Path   43
     - 3.1.1 Path Loss   44
   - 3.2 Temporal Fading   45
     - 3.2.1 Multi-path Intensity Profile   46

		3.2.2	Space-frequency Correlation Function	49

	3.2.2	Space-frequency Correlation Function	49
	3.2.3	Doppler Power Spectrum	50
	3.2.4	Space-time Correlation Function	52
	3.2.5	Frequency-selective Fading	54
	3.2.6	Flat Fading	54
	3.2.7	Fast Fading	55
	3.2.8	Slow Fading	55
3.3	Scattering Environment		55
	3.3.1	High-rank Channels	56
	3.3.2	Low-rank Channels	56
3.4	Angular Subscriber Distribution		59
3.5	Summary		61

4. SPACE-TIME CHANNEL MODELS — 63

4.1	Basic Modeling		63
4.2	Local Scattering Environment		64
	4.2.1	Circular Disk of Scatterers Model	65
		4.2.1.1 Correlation	68
	4.2.2	Gaussian Scatterer Model	69
		4.2.2.1 Correlation	74
	4.2.3	Effective Scatterer Model	75
	4.2.4	Discrete Uniform Distribution Model	76
	4.2.5	Geometrically Based Single-Bounce Statistical Models	77
		4.2.5.1 Geometrically Based Circular Macro-cell Model	78
		4.2.5.2 Geometrically Based Elliptical Micro-cell Wideband Model	78
	4.2.6	Gaussian Wide Sense Stationary Uncorrelated Scattering Model	79
	4.2.7	Raleigh's Model	81
	4.2.8	Uniform Sectored Distribution Model	82
	4.2.9	Modified Saleh-Valenzuel's Model	83
	4.2.10	Extended Tap-delay-line Model	84
	4.2.11	Measurement-based Channel Models	84
	4.2.12	Ray Tracing Models	85
4.3	Fading Distribution Based on Scattering Environment		86
	4.3.1	Exponential Fading Distribution	87
	4.3.2	Gaussian Fading Distribution	87
4.4	Summary of Space-time Channel Models		87
4.5	Summary		91

5. SMART ANTENNA TECHNIQUES — 93

5.1	Beamforming		96
	5.1.1	Fixed Beamforming	96
	5.1.2	One-shot Beamforming	97
	5.1.3	Adaptive Beamforming Algorithms	98
		5.1.3.1 Non-blind beamforming	98
		5.1.3.2 Blind Beamforming	99
		5.1.3.3 Constant Modulus Algorithm	99

			Contents	vii

		5.1.3.4	Algorithms based on DOA Estimation	101
		5.1.3.5	Algorithms based on Property-restoral Techniques	101
		5.1.3.6	Algorithms based on Discrete Signal Structure	102
		5.1.3.7	General Algorithms	102
		5.1.3.8	Integrated Algorithms for CDMA Signalling	103
	5.1.4		Correlation Influence on Beam Pattern	107
5.2	Diversity			111
	5.2.1		Transmit Diversity	111
		5.2.1.1	Code-Division Transmit Diversity	113
		5.2.1.2	Time-Division Transmit Diversity	115
		5.2.1.3	CDTD and TDTD with Pre-RAKE Combining	116
	5.2.2		Receive Diversity	117
5.3	Sectorization			120
5.4	Switched-beam Smart Antennas			120
5.5	Summary			123

6. **SMART ANTENNA PERFORMANCE** — 125

6.1	Beamforming Array Performance			125
	6.1.1	Optimum Antenna Element Spacing		129
	6.1.2	Numerical Results		133
6.2	Receive Diversity Performance			137
	6.2.1	Numerical Results		139
6.3	Combined Diversity and Beamforming Performance			140
	6.3.1	Numerical Results		140
6.4	Choosing a Spatial Processing Technique			142
6.5	Multiuser Modulation Schemes			147
6.6	Summary			151

7. **MULTIUSER DETECTION** — 153

7.1	Multiuser Detection			154
7.2	System Model for Detection			155
7.3	One-Shot Detection Techniques			157
	7.3.1	Conventional Detection		157
	7.3.2	Optimal Detection		160
	7.3.3	Linear Detection		162
		7.3.3.1	Decorrelating Detector	162
		7.3.3.2	MMSE Detector	163
		7.3.3.3	Adaptive MMSE Detection	165
	7.3.4	Decorrelating Decision Feedback Detector		167
7.4	Interference Cancellation Techniques			167
	7.4.1	Linear Interference Cancellation		170
		7.4.1.1	Jacobi Inner Iteration	171
		7.4.1.2	Gauss-Seidel Inner Iteration	172
	7.4.2	Non-linear Interference Cancellation		173
7.5	Joint Decoding for Coded CDMA			173
7.6	Numerical Examples for Interference Cancellation			175
	7.6.1	Interference Cancellation over an AWGN channel		176

	7.6.2	Interference Cancellation over UMTS Multi-path Channel Model	179
7.7	Summary		181

8. SPACE-TIME CODED TRANSMIT DIVERSITY FOR CDMA — 183
- 8.1 Space-time Coded Multiple Access — 184
- 8.2 Turbo Coding — 187
  - 8.2.1 Turbo Encoding — 188
  - 8.2.2 Turbo Decoding — 188
  - 8.2.3 Turbo Code Performance — 189
    - 8.2.3.1 Turbo Interleaver/Permuter — 189
    - 8.2.3.2 Constituent Encoder — 191
    - 8.2.3.3 Puncturing — 191
    - 8.2.3.4 Decoding Algorithm — 191
- 8.3 Space-time Coded System Model — 191
- 8.4 Layered Space-time Codes — 193
  - 8.4.1 Low Rate Code Extension — 195
- 8.5 Performance Evaluation — 196
  - 8.5.1 Convolutional Code Bounds — 196
    - 8.5.1.1 Evaluation of $P_d(\mathbf{c} \to \hat{\mathbf{c}})$ — 198
  - 8.5.2 Turbo Code Bounds — 201
- 8.6 Analytical Results — 202
  - 8.6.1 AWGN Performance — 202
  - 8.6.2 Fading Channel Multiuser Performance — 204
  - 8.6.3 CDTD and TDTD Comparison — 206
  - 8.6.4 Effects of Fading Correlation — 206
- 8.7 Simulation Performance — 211
- 8.8 Space-time Code Extensions — 214
  - 8.8.1 Turbo Transmit Diversity — 214
    - 8.8.1.1 Rate-$1/(Z+1)$ Turbo Encoding — 214
    - 8.8.1.2 Puncturing and Multiplexing — 215
    - 8.8.1.3 Iterative Decoding — 216
  - 8.8.2 Super-orthogonal Turbo Transmit Diversity — 218
- 8.9 Summary — 220

Appendices — 221
A– List of Abbreviations — 221
B– List of Symbols — 227
C– Typical Power Delay Profiles — 235
D– Correlated Multivariate Gamma Distribution — 237
- D.1 Correlated Multivariate Gamma Distribution — 237
  - D.1.1 Example — 239
  - D.1.2 Special Case — 240

E– Turbo Code Input-Output CPDF — 243
- E.1 Input-Output Weight Enumerator Recursion — 243
- E.2 Input-Output Conditional Probability Density Function — 244

| | | | Contents | ix |

F– Turbo Decoding — 247
F.1 Turbo Decoding — 247
- F.1.1 Extrinsic and Intrinsic Information — 248
  - F.1.1.1 Extrinsic Information — 248
  - F.1.1.2 Intrinsic Information — 248
- F.1.2 Soft-Input/Soft-Output Decoding Algorithm — 248
F.2 Iterative Decoding Algorithms — 250
- F.2.1 Maximum A Posteriori Algorithm — 250
  - F.2.1.1 Calculation of Branch-Metric, γ — 251
  - F.2.1.2 Calculation of Forward Recursion, α and Backward Recursion, β — 252
  - F.2.1.3 LLR Calculation, Λ — 253
  - F.2.1.4 Tail Termination — 254

G– WCDMA Simulation Environment: Physical Layer — 257
G.1 Uplink Description Summary — 258
G.2 Downlink Description Summary — 261

H– WCDMA Simulation Environment — 263
H.1 Link Level Simulation — 263
- H.1.1 Monte-Carlo Simulation Technique — 266
- H.1.2 General Simulation Assumptions — 266
- H.1.3 Simulation Cases — 267
H.2 MATLAB Simulation Software — 268
- H.2.1 Getting Started — 268
- H.2.2 Main Simulation Window — 268
- H.2.3 Simulation Environment Configuration — 268
- H.2.4 Example — 271

References — 273

Index — 299

About the Authors — 307

Disclaimer — 309

List of Figures

1.1	IMT-2000 vision.	3
1.2	IMT-2000 spectrum allocation.	4
1.3	Required spectrum for terrestrial UMTS services (including second generation services).	6
1.4	What is UMTS?	7
2.1	Cellular re-use concept.	19
2.2	SDMA system implemented using adaptive antenna arrays.	20
2.3	Space Division Multiple Access (SDMA) - allowing users in the same cell to share time/frequency and code resources.	21
2.4	Micro-, macro-, pico and hierarchical cells work together to provide seamless service.	22
2.5	An M_B-element adaptive array with a beamforming and direction finding processor.	23
2.6	$M_B \times N_B$ element planar array.	24
2.7	Some common array geometries.	26
2.8	Normalized array response as a function of angle of a ULA with $M_B = 3$ and uniform user distribution ('+' indicates user).	29
2.9	Basic block diagram of a multiple antenna CDMA transmitter.	31
2.10	Basic block diagram of a multiple antenna CDMA receiver.	32
2.11	Functional block diagram of a software radio for a base station with smart antenna.	40
2.12	The total mobile powers versus the number of users.	41
3.1	GBSBEM scatterer density geometry.	47
3.2	$S_M(\tau, \phi_r)$ as a function of DOA and TOA at base station for the GBSBEM model.	48
3.3	$|R_M(\Delta f, \phi_r)|$ as a function of DOA and TOA at base station for the GBSBEM model.	50
3.4	$S_D(f, \phi_r)$ as a function of DOA and TOA at base station for the GBSBEM model.	52
3.5	$R(\Delta t, \phi_r)$ as a function of DOA and TOA at base station for the GBSBEM model.	53

3.6	Unfolded pdf of received signal power as a function of ϕ_r.	56
3.7	Channel models for high-rank channels.	57
3.8	Channel models for low-rank channels.	58
3.9	Typical practical implementation scenario of a cellular network.	59
3.10	Qualitative description of the pdf of the mobile distributions in a Typical Urban micro-cellular scenario.	61
3.11	Example pdf of the angular distribution of mobiles.	62
4.1	Circular disk of scatterers model geometry.	66
4.2	DOA as a function of ϕ and v.	67
4.3	CDSM fading correlation for $\phi_0 = 45°$.	69
4.4	Modeling of scattering elements using the GS model.	70
4.5	Envelope correlation for rural macro- and urban micro-cells.	73
4.6	Distribution of the DOA of signals at the base station using rule of thumb calculation.	74
4.7	DOA as a function of ϕ and v.	75
4.8	GS model fading correlation.	75
4.9	ESM geometry.	76
4.10	DUD model geometry.	77
4.11	GWSSUS geometry.	80
4.12	Signal environment for Raleigh's model.	82
4.13	USD geometry.	83
4.14	Fading parameters of multi-path echoes received at the base station ($\delta_m = 1, \sigma_m = 0.6$).	88
4.15	Constructing a typical cellular channel model.	89
5.1	Classification of smart antenna techniques.	95
5.2	LS-DRMTCMA ULA beampattern for reference user located at $60°$ and interfering users at $0°$ and $-60°$ in AWGN.	105
5.3	LS-DRMTCMA ULA beampattern for reference user located at $0°$ and interfering users at $60°$ and $-60°$ in AWGN.	105
5.4	LS-DRMTCMA ULA beampattern for reference user located at $-60°$ and interfering users at $0°$ and $60°$ in AWGN.	106
5.5	LS-DRMTCMA ULA beampattern in two-path Rayleigh with reference user located at $-10°$ and multi-path signal at $0°$.	106
5.6	ULA beam pattern with envelope correlation of $\rho_{ij} = 0.8$.	108
5.7	ULA beam pattern with envelope correlation of $\rho_{ij} = 0.3$.	108
5.8	Circular array of $M_B = 12$ directional radiating elements.	109
5.9	Radiation patterns for different scan angles for the $M_B = 12$ element circular array.	109
5.10	Envelope for circular array with NLOS channel model correlation.	110
5.11	Envelope for circular array with LOS channel model correlation.	110
5.12	Envelope for circular array with macro-cell channel model correlation.	111
5.13	Block diagram of a general CDTD single user system.	113

LIST OF FIGURES   xiii

5.14	Single user CDTD receiver block diagram: a) O-CDTD b) NO-CDTD.	114
5.15	Block diagram of the space-time coded time-division transmit diversity (TDTD) system.	116
5.16	Block diagram of the pre-RAKE combining scheme for TDTD.	117
5.17	Basic structure of a diversity system.	118
5.18	Performance of MRC diversity system.	121
5.19	Sectorization and switched-beam smart antenna systems.	121
6.1	Influence of antenna spacing on total interference received by the base station for uniform user distribution (physical spacing, $d_x = \lambda/d_b$).	131
6.2	Influence of antenna spacing on total interference received by the base station for users clustered at 180° (physical spacing, $d_x = \lambda/d_b$).	131
6.3	Influence of antenna spacing on total interference received by the base station for users clustered at 90° (physical spacing, $d_x = \lambda/d_b$).	132
6.4	Influence of antenna spacing on the radiation pattern of a ULA (physical spacing as a fraction of λ given).	133
6.5	Influence of antenna array size on the BER of a cellular CDMA system including beamforming.	134
6.6	Influence of antenna array size on the capacity of a cellular CDMA system including beamforming.	135
6.7	Influence of temporal fading on the BER of a cellular CDMA system including beamforming ($M_B = 3$).	135
6.8	Influence of temporal fading on the capacity of a cellular CDMA system including beamforming ($M_B = 3$).	136
6.9	Composition of beamforming and diversity signals.	137
6.10	BER of diversity system under LOS conditions.	140
6.11	BER of diversity system under NLOS conditions.	141
6.12	Capacity of diversity system under LOS and NLOS conditions.	141
6.13	BER of diversity system under LOS conditions.	142
6.14	BER of diversity system under NLOS conditions.	143
6.15	Capacity of diversity system under LOS and NLOS conditions.	143
6.16	A comparison of $M_D = 3$ antenna diversity and $M_B = 3$ element beamforming systems under LOS conditions.	144
6.17	A comparison of $M_D = 3$ antenna diversity and $M_B = 3$ element beamforming systems under NLOS conditions.	145
6.18	A comparison of a combined $M_B = 3$ element antenna beamformer with $M_D = 3$ diversity branches and a $M_B = 9$ element beamforming systems under LOS conditions.	146
6.19	A comparison of a combined $M_B = 3$ element antenna beamformer with three diversity branches and a $M_B = 9$ element beamforming systems under NLOS conditions.	147

6.20	Probability of error in two-user BPSK and QPSK system with $N = 16$.	148
6.21	Probability of error in two-user BPSK and QPSK system with complex spreading sequences and $N = 16$.	149
6.22	Probability of error performance for three equivalent systems of the same bandwidth efficiency, using complex spreading codes and assuming independent one-path fading channels for each user.	150
6.23	Probability of error performance over all users at the output of the third stage of a clip-function SIC, for $K = 4, 8$ BPSK and $K = 4$ QPSK systems having identical bandwidth efficiencies (complex spreading codes and complex channels).	150
7.1	The fundamental structure of **R** where $K = 3$ and $P = 5$. The circles on the diagonal represent the autocorrelations. The remaining dots represent non-zero cross correlations. The band-diagonal structure is obvious.	158
7.2	Interference power versus desired signal power as a function of the number of active users.	160
7.3	Different mapping functions, a) soft decision, b) hard decision, c) clipped soft decision.	169
7.4	The fundamental structure of a powerful iterative receiver for joint detection and decoding.	175
7.5	The performance of the 3 SIC schemes over an AWGN channel, $K = 10$, $N = 32$. Each SIC has 3 stages. The matched filter detector and the single user bound are included for reference.	176
7.6	The performance of the 3 SIC schemes over an AWGN channel, $K = 24$, $N = 32$. Each SIC has 3 stages. The matched filter detector and the single user bound are included for reference.	177
7.7	The performance of the 3 PIC schemes over an AWGN channel, $K = 10$, $N = 32$. Each SIC has 3 stages. The matched filter detector and the single user bound are included for reference.	178
7.8	The performance of the 3 PIC schemes over an AWGN channel, $K = 24$, $N = 32$. Each SIC has 3 stages. The matched filter detector and the single user bound are included for reference.	178
7.9	Performance comparison of PIC and SIC schemes over an AWGN channel, $K = 10$, $N = 32$. The matched filter detector and the single user bound are included for reference.	179
7.10	The performance of the 3 SIC schemes over the UMTS vehicular channel model, $K = 10$, $N = 32$. Each SIC has 3 stages. The matched filter detector is included for reference.	180

7.11	Performance comparison of the MF detector and the clip SIC over the UMTS vehicular channel model, $K = 10$, $N = 32$ with (1,4,8) element antenna arrays. Each SIC has 3 stages.	180
8.1	Coded transmit diversity space.	185
8.2	Turbo coded system design space [28].	190
8.3	Block diagram of a space-time coded transmit diversity CDMA system.	192
8.4	Generalized block diagram of the space-time convolutional coder based on a sub-optimum configuration, (a) Encoder, (b) Decoder.	194
8.5	Generalized block diagram of the space-time convolutional coder based on a per-user configuration, (a) Encoder, (b) Decoder, with soft-information transfer.	195
8.6	CC and TC bounds on AWGN channel with $R_c = 1/2$ and $K = 1$, as a function of turbo interleaver size, N_{tc}.	203
8.7	CC and TC bounds on AWGN channel with $N'_{tc} = 256$ and $K = 1$, as a function of code rate.	203
8.8	Analytical AWGN system load V – low rate convolutional and turbo coding ($N_{tc} = 256$).	204
8.9	Analytical BER performance of coded O-CDTD, with $R_c = 1/2$, $K = 5$, and $M_T = 1, 2, 3$, on a fast fading 2-path channel.	205
8.10	Analytical BER performance of coded O-CDTD, with $R_c = 1/2$, $K = 5$, and $M_T = 1, 2, 3$, on a slow fading 2-path channel.	205
8.11	Coded O-CDTD BER performance on a 2-path fading channel with $R_c = 1/2$ and $M_T = 1$.	207
8.12	Coded O-CDTD BER performance on a 2-path fading channel with $R_c = 1/2$ and $M_T = 2$.	207
8.13	Coded O-CDTD BER performance on a 2-path fading channel with $N_{tc} = 256$ and $M_T = 1$.	208
8.14	Coded O-CDTD BER performance on a 2-path fading channel with $N_{tc} = 256$ and $M_T = 2$.	208
8.15	Coded O-CDTD performance comparison with $R_c = 1/2$, $N_{tc} = 256, 2048$ and $M_T = 1, 3$.	209
8.16	Comparison of $R_c = 1/2$ convolutional and turbo coding ($N_{tc} = 256, 2048$) CDTD and AS-TDTD.	209
8.17	Comparison of $R_c = 1/2$ convolutional and turbo coding ($N_{tc} = 256, 2048$), with $M_T = 3$ and $\rho = 0, 0.5, 0.99$.	210
8.18	O-CDTD NLOS performance with $M_T = 3, 5$.	211
8.19	Average error rate for O-CDTD in the WCDMA vehicular environment (120 km/h).	212
8.20	Average error rate for different transmit diversity schemes in the WCDMA vehicular environment (120 km/h).	212

8.21	Generalized TTD block diagram.	215
8.22	Puncturing and multiplexing procedure for a rate $R_c = 1/2$ turbo encoder with Z constituent RSC encoders. (a) Single transmit antenna, $M_T = 1$ (b) $M_T = 3$ transmit antennas.	217
8.23	Generalized block diagram of a serial iterative turbo decoder.	218
8.24	SOTTD block diagram.	219
D.1	Probability density function for $M = 4$ and $m = 4$	240
E.1	Conditional probability density function for output weight d of constituent recursive convolutional encoder given input weight i.	245
F.1	Soft-Input/Soft-Output (SISO) decoder for a rate-1/2 RSC code.	249
F.2	Turbo decoder block diagram.	250
F.3	Graphical illustration of the computation of the branch metric, γ.	252
F.4	Graphical illustration of the computation of the forward recursion, α.	253
F.5	Graphical illustration of the computation of the backward recursion, β.	253
G.1	Uplink spreading and modulation for DPDCH and DPCCH.	258
G.2	Frame structure for the uplink DPDCH/DPCCH channels.	259
G.3	Code generation tree for OVSF codes.	259
G.4	Frame structure for the downlink DPDCH/DPCCH channels.	261
G.5	Downlink spreading and modulation for DPDCH and DPCCH.	262
H.1	Overall block diagram of the uplink.	264
H.2	Overall block diagram of the downlink.	265
H.3	Main interactive simulation platform window.	269
H.4	Simulation platform configuration window.	271

List of Tables

1.1	Expected growth in wireless users (in millions).	2
1.2	Future terrestrial spectrum requirements.	5
1.3	Envisioned 2002 content of UMTS Phase 1.	10
1.4	Broadband service deployment options for various scenarios.	12
4.1	Cell radii and local scattering element standard deviations for macro- and micro-cellular environments.	72
4.2	Comparison of angular spread calculated using rule of thumb and typical measured angular widths.	74
4.3	Criteria for selecting τ_m.	79
4.4	Summary of temporal fading model.	88
5.1	Correlation values for different channel models for a $M_B = 12$ element circular array.	109
6.1	System parameters for numerical evaluation of BER performance.	134
6.2	System parameters for numerical evaluation of BER performance with combined beamforming and diversity.	144
8.1	System parameters for simulation performance.	213
8.2	Orthogonality factor for the different WCDMA channel environments.	213
G.1	Key parameters for UMTS.	257
H.1	Simulation service classes.	267
H.2	Summary of simulation features.	267

Preface

The field of smart antenna technology or more generally, space-time processing, is rapidly becoming one of the most promising areas for the improvement of the capacity of mobile communication systems. It is especially in the context of the development of the first practical third generation mobile communication systems that space-time processing is attracting increased interest from both academic and industrial circles. It is with this background in mind that this book has been written. Specifically, we have tried to answer some of the most basic questions relating to the use of space-time processing in CDMA based third generation mobile communication systems, in addition to presenting some models for the integration of space-time processing, error correction coding and multiuser detection techniques.

The book is aimed at two groups of readers. First, we have tried to include all the necessary background information and analytic tools required by research students and engineers to become proficient in the analysis of space-time processing systems. Furthermore, the book presents comprehensive introductions into the topics of space-time coding and multiuser detection. As the main focus of the book is on space-time processing techniques, these topics are presented in the context of designing a mobile communication systems where the three core areas of spatial processing, coding and multiuser detection is to be integrated in an optimum way.

Secondly, the book will prove to be a valuable guide to design engineers enabling them to understand the implications of adding space-time processing systems to CDMA based communication systems. Specifically, the book will help design engineers to make decisions as to where and when which type of space-time processing technique will yield the largest gains in system performance. Furthermore, several issues relating to the implementation of smart antenna systems are discussed in detail.

In covering the vast topic of space-time processing, this books concentrates mainly on three aspects. In Chapters 2 to 4, extensive background information on cellular systems, basic antenna array theory, channel impairments and space-time channel models are presented. Together these chapters give the reader all the necessary background information required to understand specific smart antenna techniques as well as the derivation of the BER performance of these techniques in cellular environments. Chapter 5 focuses specifically on smart antenna techniques. The four main smart antenna techniques covered are adaptive beamforming systems, transmit and receive diversity systems, sectorization and switched beam antennas. With this background mastered, the reader should be in a position to derive the theoretical performance of various space-time processing systems as used in cellular environments. Thus, in Chapter 6 the performance of basic space-time processors such as beamformers, diversity and combined diversity and beamforming systems are derived as an introduction into the theoretical evaluation of space-time processors. These concepts are then extended in Chapters 7 and 8 to include advanced space-time processors utilizing multiuser detection and space-time coding.

The book is underpinned by an extensive simulation program, called the WCDMA Simulation Environment, written in Matlab®. The simulation code is available with the book, as well as on the Matlab® web site and implements both the uplink and the downlink of a UMTS-like communication system. The reader is provided the option to simulate the system performance using a variety of channel models as well as space-time receiver structures. For example, the user can select an UMTS Indoor Channel with transmit and receive diversity, turbo coding and multiuser detection in one simulation. Clearly this is a very powerful tool and the reader is encouraged to use this simulation package to explore the theoretical concepts discussed in the book and space-time concepts in general.

Our last words in this preface must be ones of thanks as many people contributed to making this book a reality. Firstly, we would like to thank Lars Rasmussen for his contribution on multiuser detection (Chapter 7). In addition to contributing this chapter, we would also like to thank him for his valuable comments on the structure of the book. Also, we need to thank various colleagues at the University of Pretoria, the University of South Africa, Defencetek and various other institutions for their contributions. Specifically, we acknowledge contributions from Johan Joubert, Wimpie Odendaal, Eugene Botha, Vladimir Katkovnik, Sharma Yadavalli, Paul Alexander (who contributed a large portion of the simulation software), Matti Latva-aho, Jack Glass, Giorgio Taricco and the students of the Alcatel Research Unit for Wireless Access at the University of Pretoria. A special word of thanks to the University of Pretoria, Alcatel, Altech and the CSIR Defencetek for allowing us to complete this project. A lot of credit also needs to go to Alex Greene and the staff at Kluwer for their encouragement and support of this project. Last, but definitely not least, we would like to thank our families and friends for putting up with us during the completion of the manuscript - without their support it would not have been possible!

PvR ML DvW

... that is what learning is. You suddenly understand something you've understood all your life, but in a new way.
Doris Lessing

1 INTRODUCTION

Moore's Law: *Transistor density on integrated circuits doubles every 18 months.*

Metcalfe's Law: *A new communications application will probably be stillborn because the initial value will be so small that no one will have sufficient incentive for purchase.*

The challenge to any new telecommunication system is to capitalize on Moore's Law and to buck Metcalfe's Law. Two systems have managed to achieve this awe-inspiring feat in the past five to ten years, namely the Internet and mobile communications. Both have capitalized on the increased computing power available from general purposes and application specific processors, and both have managed to overcome the "critical mass" entry barrier described by Metcalfe's law.

Moving from relative obscurity only four to five years ago to being one of the world's most important communication and commerce platforms, the Internet is revolutionizing wireline communications. In addition to the social changes that it is inducing, it is shaping the future development of technologies such as integrated services digital network (ISDN), asymmetric digital subscriber loops (ADSL), asynchronous transfer mode (ATM) and also mobile communications [163]. Estimates of the demand for broad band services for fixed and mobile applications vary substantially, but most analysts agree that users of e-mail, file transfer protocol and other Internet services will demand increased access to bandwidth [168].

2 SPACE-TIME PROCESSING FOR CDMA

Year	Europe	North America	Asia Pacific	Rest of the world	Total
2000	113	127	149	37	426
2005	200	190	400	150	940
2010	260	220	850	400	1730

Table 1.1. Expected growth in wireless users (in millions) [221].

In addition to the explosion of the Internet, mobile communication is one of the fastest growing industries in the world. This growth phenomenon is exemplified by the near exponential growth in the number of cellular subscribers in Europe and Japan [50] (see Table 1.1) and also by the fact that six of the world's top ten companies in the electronics sector are involved in the wireless access business in some form or another [156].

1.1 IMT-2000

Current development trends in telecommunications are driven by user requirements, which include access to a diverse range of services for anyone, anywhere, anytime and at the lowest possible cost. In an ideal world, this would ultimately lead to one, worldwide mobile solution, using one radio-access network and one single core network. However, the trend towards a deregulated worldwide market and the rapid introduction of incompatible second generation mobile services makes this an unrealistic dream. The approach to achieving the stated user requirements is, therefore, for the existing networks and services to evolve at their own pace towards the common goal of a global telecommunications standard. This influences the current standardization process for global telecommunications at national, regional and International Telecommunications Union (ITU) levels. Such a flexible approach opens possibilities for active collaboration between national and regional standardization bodies and the ITU. At the same time, it offers possibilities for all players to continue developing at their own speed.

International Mobile Telecommunications for the 21st century (IMT-2000), initially known as the Future Personal Land Mobile Telephone System (FP-LMTS), is an initiative of the ITU aimed at providing wireless access to the global telecommunication infrastructure through both satellite and terrestrial systems serving fixed and mobile users in public and private networks. It is being developed on the basis of the "family of systems" concept, providing IMT-2000 service capabilities to users of all family members in a global roaming offering. The ITU vision of global wireless access in the 21st century is aimed at providing direction for the many related technological developments in this area so as to assist in the convergence of these essentially competing wireless access technologies. Figure 1.1 graphically illustrates this vision.

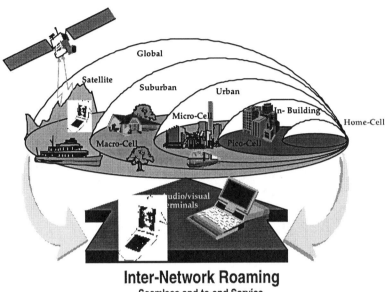

Inter-Network Roaming
Seamless end-to-end Service

Figure 1.1. IMT-2000 vision [221].

The role of the ITU in the development of IMT-2000 is to ensure recognition of the IMT-2000 family concept, to register IMT-2000 family members and to facilitate the important criteria which enable roaming between members of the IMT-2000 family. The ITU, therefore, has to facilitate negotiations on issues such as service inter-working, framework standardization and spectrum compatibility. To achieve these goals, the ITU has to seek commonalities between IMT-2000 systems and ensure free circulation of information between all participating bodies.

Recently the various standardization bodies recognized the different market needs in different regions and called for evolution paths from all second generation systems and inter-standard inter-working between second generation systems to pave the way for global roaming. Global standards should, therefore, address the kernel of services and interfaces necessary for international compatibility whereas regional standardization organizations should specify the details and co-operate on service inter-working. Since market needs and interests are different between regions, it should be recognized that there will be different IMT-2000 systems based on a common framework.

1.2 IMT-2000 SPECTRUM ALLOCATION

Adequate access to spectrum is a key requirement for the development of IMT-2000. In 1992, the World Radio Conference identified the frequency bands 1885 - 2025 MHz and 2110-2200 MHz for future IMT-2000 systems. Of these the bands 1980 - 2010 MHz and 2170 - 2200 MHz were intended for the satellite part of such future systems.

Europe and Japan have decided to implement the terrestrial part of UMTS (the UMTS Terrestrial Radio Access, UTRA, air interface) in the paired bands 1920 - 1980 MHz and 2110 - 2170 MHz. Europe has also decided to implement UTRA in the unpaired bands 1900 - 1920 MHz and 2010 - 2025 MHz. In early 1998, the European Commission published the "EC Proposal for a European Parliament and Council decision on the Co-ordinated Introduction of UMTS" to ensure that EU member states undertake the appropriate steps needed to implement the European Radio Committee's (ERC) decision on spectrum assignment. This, in combination with the existing Licensing Directive, will ensure UMTS services can commence in 2002.

Figure 1.2 indicates the spectrum allocations of each of the major users of IMT-2000. It should be clear that the spectrum demand for IMT-2000 is significantly dependent on the number of users, the quantity of traffic flow and its distribution.

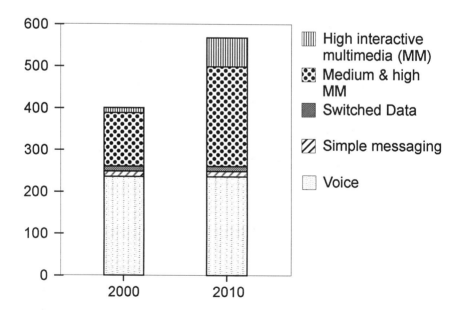

Figure 1.2. IMT-2000 spectrum allocation [221].

Table 1.2 and Figure 1.3 show the required frequencies per service in typical busy hours for the years 2005 and 2010. The conclusion is that roughly 580 MHz will be required in the year 2010. The requirement includes the bands currently designated for second generation systems, and the bands designated as core bands for UMTS, plus new spectrum resources fully and flexibly exploited. It is envisaged that the increase in penetration after 2010 will not be significant [221]. The use of services requiring wider bandwidth, however, is expected to increase, which will lead to increasing spectrum demand.

Year	2005	2010
High interactive multimedia[1]	22 MHz	82 MHz
Medium and high multimedia[2]	113 MHz	241 MHz
Switched data	12 MHz	9 MHz
Simple messaging	2 MHz	2 MHz
Voice	220 MHz	220 MHz
Total	369 MHz[3]	554 MHz[4]
Total (allowing for spectrum division)	406 MHz	582 MHz

Table 1.2. Future terrestrial spectrum requirements [221].

1.3 FROM SECOND GENERATION TO THIRD GENERATION

A good example of the migration from second generation mobile systems (GSM, IS-95, IS-54, etc) to the IMT-2000 vision (i.e. UMTS), is the evolution from the European Telecommunications Standardization Institute (ETSI) defined GSM system to UMTS. The UMTS system is only one of many new third generation systems being developed around the world, and serves as illustration for our current discussion.

The success of the European second generation system, GSM, has created a mass market for mobile communications, reaching high terminal penetration in global markets. At the end of June 1998, there were 293 members of the GSM memorandum of understanding (MoU) association from 120 different countries worldwide. There are currently 278 GSM networks in operation serving 95 million subscribers and these are still growing. It is expected that 235 million subscribers will be served by the end of the year 2000. A further boost to the mass market will be the introduction of multi-mode multi-band terminals, such as GSM/DCS 1800/PCS 1900, GSM/satellite and many other handset combinations. The expected penetration for mobile communications in developed countries is expected to rise to 50%-80% within the time frame from the introduction of UMTS.

UMTS cannot be developed as a completely isolated network with minimal interface and service interconnection to existing networks. Both UMTS and

6 SPACE-TIME PROCESSING FOR CDMA

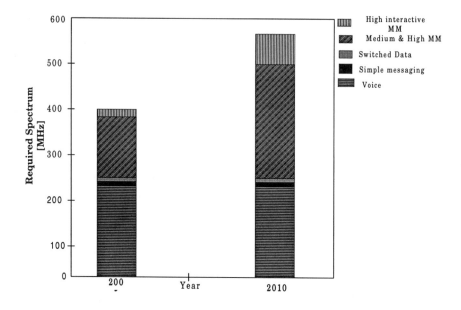

Figure 1.3. Required spectrum for terrestrial UMTS services (including second generation services) [221].

existing networks will need to develop along parallel, even convergent paths, if service transparency is to be achieved to any degree. This would then, in the end, allow UMTS services to be supported, although at different levels of functionality, across all networks. Another important requirement for seamless operation of the two standards is GSM-UMTS hand-over in both directions. Restrictions on the applicability of hand-over may be necessary for particular services and when services are different between the systems. This will require modifications of existing GSM specifications. GSM networks also need to be protected from unwanted side-effects caused by functions needed to support cross hand-overs.

It should be noted that this migration approach will enable a full exploitation of current systems. Moreover, due to the significant investments made, pre-UMTS operators will not want to discard their existing infrastructure and will prefer some type of co-existence and interworking between existing and new infrastructure. The balance between use of the current systems and UMTS systems will ultimately be decided by market forces.

1.4 UNIVERSAL MOBILE TELECOMMUNICATIONS SYSTEM

UMTS wideband code division multiple access (WCDMA) is one of the major new third generation mobile communication systems being developed within the IMT-2000 framework. The subject of intense worldwide research and development efforts throughout this decade, UMTS has the support of many major telecommunications operators and manufacturers because it represents a unique opportunity to create a mass market for highly personalized and user-friendly mobile access to tomorrow's "Information Society".

It represents a substantial advance over existing mobile communications systems. Above all else it is being designed with flexibility for users, network operators and service developers in mind and embodies many new and different concepts and technologies. Figure 1.4 attempts to summarize some of the key elements and features which, in combination, seek to define the scope of UMTS.

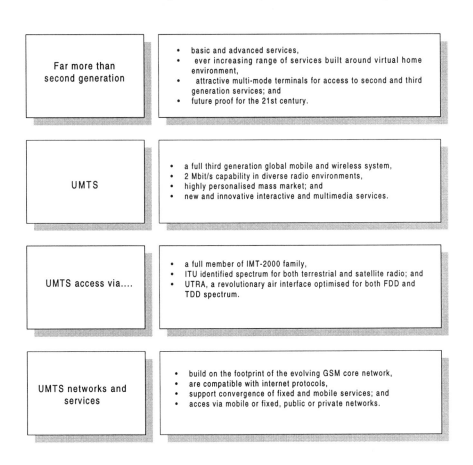

Figure 1.4.   What is UMTS?

UMTS seeks to build on and extend the capabilities of today's mobile, cordless and satellite technologies by providing increased capacity, data capability and a far greater range of services using an innovative radio access scheme and an enhanced, evolving core network.

One factor which sets UMTS clearly apart from second generation mobile systems is its potential to support 2 Mbit/s data rates from the outset. This capacity, together with inherent Internet protocol (IP) support, provides a powerful combination for delivering interactive multimedia (MM) services as well as other new wideband applications such as video telephony and video conferencing. As the demand for user data rates increases in the long term, UMTS will be developed to support even higher data rates, perhaps one or two orders of magnitude greater (provided appropriate spectrum is allocated). In later phases of UMTS development there will be a convergence with even higher data rate systems using mobile wireless local area network (LAN) technologies (microwave or infrared) providing data rates of, for example, 155 Mbit/s in indoor environments.

Most cellular systems in use today rely on circuit switched technology for wireless data transmission. However, UMTS integrates packet and circuit data transmission, since packet data over-the-air affords several benefits to users. These are

- virtual connectivity to the network at all times;

- alternative ways of billing - for example pay-per-bit, per session or flat monthly rate;

- asymmetric bandwidth in the uplink and downlink - as demanded by many emerging data services where one link direction carries simple commands and the other carries the content rich, bandwidth intensive traffic (for example web browsing or video transmission).

UMTS is also being designed to offer data rate on demand, where the network reacts flexibly to the user's demands, the customer's profile and the current status of the network. The use of packet-oriented transport protocols such as IP is being studied so that UMTS can enhance these abilities. Together, the combination of packet data and data rate on demand will remove technical barriers for the user and make operation of the system much cheaper - there will be no worries about how and when to connect to the network.

UMTS services are based on standardized service capabilities which are common throughout all UMTS user and radio environments. This means that a personal user will experience a consistent set of services even when he roams from his home network to other UMTS operators - a "virtual home environment" (VHE). Users will always "feel" that they are connected to their home network, even when roaming. VHEs will ensure the delivery of the service provider's total environment, including, for example, a corporate user's virtual work environment, independent of the user's location or mode of access (satellite or terrestrial). VHEs will also enable terminals to negotiate functionality with the visited network, possibly even downloading software in order to provide "home-like" services with full security and transparency across a mix of access and core networks. The ultimate goal is that all networks, signalling, connections, registrations and any other technologies should be invisible to the user, ensuring that mobile multimedia services are simple, user-friendly and effective.

The UMTS terrestrial radio access (UTRA) system will support operation with high spectral efficiency and service quality in all the physical environments in which wireless and mobile communication take place. Today's user lives in a multi-dimensional world, moving between indoor and outdoor congested (urban), and outdoor rural environments with mobility ranging from essentially stationary through pedestrian up to very high speeds. There are also different user density environments, including three dimensional situations in high rise buildings. UTRA is being specified to operate in all these environments. In practical implementations of UMTS, some users may be unable to access the highest data rates at all times. For example, the physical constraints of radio propagation and the economics of operating a network will mean that the system services might only support lower data rates in remote or heavily congested areas. Therefore, in order to ensure that the subscriber is always able to use his or her terminal, services will be adaptive to different data rate availability and other quality of service (QoS) parameters.

The key features of UMTS discussed thus far have been approved by ETSI SMG and are the basis for the development of standards for UMTS Phase 1. These features are summarized in Table 1.3.

1.5 THIRD GENERATION AND BEYOND

In order to ensure that the IMT-2000 vision is sustained in the long term, its capabilities will need to be progressively increased by the addition of new technologies. A selection of these are detailed below.

1.5.1 Re-configurable Terminals

Future mobile terminals will have to exist in a world of multiple standards – both second generation and those of other members of the IMT-2000 family. Also, standards themselves are expected to evolve. In order to provide universal coverage, seamless roaming and non-standardized services, some of the elements of the radio interface (e.g. channel coder, modulator, transcoder, etc.) will no longer have fixed parameters, rather they will take the form of a "toolbox" whereby key parameters can be selected or negotiated to match the requirements of the local radio channel. In addition to the ability to adapt to different standards described above, downloadable terminals will enable network operators to distribute new communications software over the air in order to improve the terminals' performance in the network or to fix minor problems. An example might be the downloading of an improved hand-over algorithm. This aspect of software download will generally be invisible to the user.

Services	Multimedia services: • $\leq$ 2 Mbit/s; $\leq$ 10 km/h • suburban; $\leq$ 384 kbit/s; $\leq$ 120 km/h • rural; $\leq$ 144 kbit/s; $\leq$ 500 km/h Low bit rate, high quality speech (like fixed networks) Packet & circuit switched services for different data rates and different radio environments Service creation and measurement toolkit Service portability when roaming into other networks Advanced addressing mechanism, e.g. personal, Internet-style New charging mechanisms (e.g. volume) Dual band/mode of operation UMTS/GSM, including roaming between UMTS "islands" Operation in any suitable band that becomes available, e.g. GSM, DCS 1800, PCS 1900
Terminals	Mobiles and SIM with downloading capabilities over the air for e.g. data and applications (feasibility in Phase 1 needs further study) Multimedia terminals Dual mode/band GSM/UMTS terminals Adaptive terminals
Access network	New UMTS BSS (Base station subsystem): • flexible bearer • rates $\leq$ 2 Mbit/s • fast, self adapting interface High capacity Support of variable bit rate and of mixed traffic types High spectrum efficiency for multimedia and low bit rate speech
Core transport network	Evolution of GSM NSS (Network switching subsystem) and ISDN/IN (Intelligent network) New charging and accounting mechanisms Support of service mobility across networks - VHE (Virtual home network) Support of variable bit rates and mixed traffic types Mobile fixed convergence elements Support of packet data by Internet protocols
Security	Protection of network use Provision of security services to the user Control of misuse and/or abuse of the network
Operation & Maintenance	Automatic establishment of roaming relations Support of multi-vendor networks

Table 1.3. Envisioned 2002 content of UMTS Phase 1 [66].

1.5.2 Application and Service Download

In using today's multimedia terminals (for example PCs), users have learned to accept the idea that the capabilities of the terminal can be modified over time by software downloading. It is now commonplace for a user to download a new "plug in" (for example a video or audio codec) to access new types of content. The introduction of multimedia services in UMTS will take this concept into the mobile domain. UMTS "plug ins" will come from a variety of sources. For example, they may be

- pre-installed on the user's terminal by the network operator or service provider,

- downloaded over the air, at the user's request or automatically by the network, much as today, where many Internet service providers upgrade one's software or databases during a session, or

- supplied on media such as CD-ROM, which may be distributed free with magazines or by direct mail.

This concept of software download will be closely linked with newly developed subscriber identity module (SIM) capabilities. The terminal and the SIM will co-operate in requesting, storing and executing software plug ins - ideally the majority of new software should be stored on the SIM to allow the user to "SIM roam" onto a new terminal while still keeping the optimum home environment.

1.5.3 Broadband Satellite Systems

Several broadband satellite systems are also planned for deployment in the post 2002 time frame (designed to offer data rates beyond 2 Mbit/s and into the Gigabits domain). Some of these systems may offer compatibility with UMTS service concepts using satellite frequency allocations in the 20/30 GHz range. The requirements of the terminal equipment and higher power consumption will necessitate larger transportable or fixed terminals.

1.5.4 Space-time Techniques

Space-time techniques are a key way to enhance the capability of mobile communication services in the long term, and are currently regarded by many within the wireless communications industry as a core system component in future-generation mobile networks. For example, the current UMTS standard already provides for antenna array use. The pilot bits available in the dedicated physical channels ensure that space-time technology can be introduced in the future. Thus, in order to promote European research and development in this strategic area, funding has been made available, through the RACE and ACTS programs, to the TSUNAMI consortium. For an in depth discussion of the TSUNAMI project, [276] should be consulted. As an example of a space-time

12   SPACE-TIME PROCESSING FOR CDMA

processing technique, antenna arrays react intelligently to the received radio signal, continually modifying their parameters to optimize the transmitted and received signal. This allows them to

- increase coverage and capacity by reducing interference between adjacent mobiles;

- offer space division multiple access (SDMA), where frequencies are assigned on a per-mobile rather than a per-cell basis allowing for vastly increased capacity; and

- enable user location in space, thus allowing the introduction of advanced location-based services.

1.6 SPACE-TIME PROCESSING SYSTEMS

Given the discussion on the introduction of UMTS as an example of a third generation mobile system, the importance and goals of this book must be seen against the backdrop of an ever growing user population that is making increasingly stringent demands for high quality, flexible services from existing and planned communication networks.

Mobility	User Type	Fiber	Copper	Radio	Satellite
Fixed	Business	APON FTTB SoNet/SDH	B-ISDN xDSL HFC	MMDS W-LAN W-ATM HFR	V-SAT MM LEO
	Residential	FTTH FTTC	xDSL HFC ISDN 56K Modem	WLL LMDS MMDS CTM	MM LEO
Mobile	Pedestrian			W-ATM UMTS IMT-2000	MM LEO Voice LEO S-UMTS
	Vehicular			2G Solutions UMTS IMT-2000	MM LEO Voice LEO

Table 1.4.   Broadband service deployment options for various scenarios.

These high quality, flexible services can be delivered to end users in a number of ways. Table 1.4 summarizes some of the possible broad band network deployment scenarios [85, 50, 185, 245, 3, 6, 78]. The increased interest in radio-based wireless access solutions shown in Table 1.4 has, however, come at a price. Increasingly, new and existing network operators have to compete for the very limited spectrum available to deliver the above mentioned services. A typical example of the high cost of this natural resource is the $17.2 billion raised by the auctioning of licenses for the delivery of PCS in the United States of America [329]. This high capital outlay has driven both network operators and equipment suppliers to develop innovative wireless access solutions that maximize the efficient use of the available spectrum or the spectral efficiency of networks. Many novel solutions in the areas of coding [329, 25, 253, 117], modulation [249, 71, 177, 313, 203], planning [314, 247] and access methodologies [141, 207, 88, 121], among others, have been proposed, studied and implemented in the recent past. Among these are the use of antenna arrays as diversity, switched beam, or beamforming systems to increase the capacity of cellular access networks. The solutions printed in *italics* in Table 1.4 are typical examples of cellular access systems that can use space-time techniques to increase the spectral efficiency of networks. Of particular importance in this book will be those CDMA-based solutions for the mobile sector and the performance analysis issues that need to be addressed as a result of the introduction of heterogeneous services and service environments into a single, mobile cellular access network.

In addition, space-time techniques do not stand on their own in a communication system. More and more, the design of a communication system requires some form of global optimization to reach the capacity targets. Thus, whereas the focus of this book is on space-time technology in general, it would not be complete without a discussion of the interaction of the antenna systems with the other main system components, viz. the receiver and the error control blocks. Furthermore, when this interaction is discussed, it is important to examine future proof techniques such as multiuser detection (MUD) and turbo coding, as they will form the basis of mobile radio systems.

1.7 ORGANIZATION OF BOOK

In covering the vast topic of space-time processing, this book concentrates mainly on three aspects. In Chapters 2 to 4, extensive background information on cellular systems, basic antenna array theory, channel impairments and space-time channel models are presented. Together these chapters give the reader all the necessary background information required to understand specific smart antenna techniques as well as the derivation of the BER performance of these techniques in cellular environments. Chapter 5 focuses specifically on smart antenna techniques. The four main smart antenna techniques covered are adaptive beamforming systems, transmit and receive diversity systems, sectorization and switched beam antennas. With this background mastered, the reader should be in a position to derive the theoretical performance of various space-time processing systems as used in cellular environments. Thus, in Chapter 6 the performance of basic space-time processors such as beamformers, diversity and combined diversity and beamforming systems are derived as an introduction into the theoretical evaluation of space-time processors. These concepts are then extended in Chapters 7 and 8 to include advanced space-time processors that include multiuser detection and space-time coding. A more detailed description of the contents of each chapter is presented below.

In Chapter 2, the basic concepts of space-time processing techniques as applied in third generation mobile communication systems are introduced. As third generation communication systems rely on cellular principles, the chapter begins by introducing the relevant cellular concepts and their relation to the use of space-time processing techniques. Specific attention is also given to aspects such as the practical implementation of space-time processing systems.

As in all communication systems, the environment in which the space-time processing system operates significantly influences the system gains that can be achieved. Therefore, in order to understand the operation of space-time processors, channel aspects influencing overall system performance are discussed in Chapter 3. Specifically, the chapter covers aspects such as propagation path, temporal fading and scattering environment.

Once the basic properties of the channel are understood, a channel model that includes all the major aspects influencing system performance can be constructed. Chapter 4 discusses a number of scattering models focusing specifically on the circular disk of scatterers model as well as the Gaussian scatterer model. Combined with a model for the fading distribution of signals in a mobile environment, the scattering model is used to construct a comprehensive space-time channel model suitable for the evaluation of space-time processing techniques.

Up to this point, space-time processing systems have been treated in general. Chapter 5 presents a detailed discussion of the basic space-time processing techniques. Specifically, detailed discussion of beamforming, transmit diversity, receive diversity, sectorization and switched beam systems are presented. Also, various general purpose beamforming algorithms, such as the LMS and CMA algorithms are discussed.

Given all the tools necessary to evaluate space-time processing systems, Chapter 6 shows the reader step-by-step how to analyze the three main types of space-time processors namely beamforming systems, diversity systems and combined beamforming and diversity systems. The chapter discusses the tradeoffs between the systems and also touches on some practical implementation aspects.

Once the reader has mastered all the necessary techniques to evaluate space-time processing systems, the book introduces the first of two advanced space-time processing techniques in Chapter 7 namely multiuser detection techniques. In this chapter, the basic principles of multiuser detection are introduced and several types of multiuser detection schemes are treated in detail. Specifically, the interaction between multiuser detection schemes and multiple antenna systems are discussed. The multiuser detection techniques discussed in this chapter are included in a comprehensive simulation of a UMTS-like CDMA system.

Finally, in Chapter 8, the concept of space-time coding for CDMA system is introduced as an example of an advanced space-time processor that could be used in a third generation mobile communication system. The chapter shows the reader how to construct space-time codes and specifically considers application using convolutional and turbo codes. The coding systems described in this chapter are also included in the WCDMA Simulation Environment described in Appendix H, and can be integrated with diversity and multiuser detection. Appendix E and F describe transfer function bounds and turbo decoding implementation.

The UMTS-like WCDMA Simulation Environment has been written in MATLAB® and the simulation code is available with the book, as well as on the Mathworks web site[5] and implements both the uplink and the downlink of a UMTS-like communication system. The reader is provided the option to simulate the system performance using a variety of channel models as well as space-time receiver structures. For example, the user can select an UMTS Indoor Channel with transmit and receive diversity, turbo coding and multiuser detection in one simulation.

Notes

1. High speed data rates, symmetric and reasonably continuous transmission and minimum delays.
2. Moderate data rates, medium to large files, asymmetric and bursty transmission and tolerance to a range of delays.
3. Already identified spectrum is 395 MHz (70 MHz GSM+150 MHz GSM 1800 + 20 MHz DECT +155 MHz terrestrial UMTS.)
4. Trunking inefficiency and guard-bands must be allowed for, due to multiple operators and public/private and service category segmentation. This is assumed to improve from 10% in 2005 to 5% in 2010.
5. http://www.mathworks.com

2 BASIC SPACE-TIME ASPECTS

The reasoning behind the use of space-time processing techniques (which include smart antennas as a special case) is the optimization of the cellular spectral efficiency of the network. This is realized by implementing more than one antenna element to optimally transmit or receive signals by using both temporal and spatial signal processing techniques in the transceiver. Well known techniques such as antenna sectorization (spatial signal processing), diversity combining (spatial and temporal signal processing) and beamforming arrays (spatial and temporal signal processing) are considered to be examples of space-time processing. In fact, all antenna array systems can be considered to be space-time processors. More advanced space-time processors include multiuser detectors, discussed in detail in Chapter 7, and space-time coding, discussed in Chapter 8, and form an integral part of space-time processing.

Implementation and performance improvements of space-time processing are, to a large extent, related to the type of system under consideration. Third generation systems based on the IMT-2000 vision (such as UMTS) have adopted wideband code division multiple access (WCDMA), Time Division CDMA (TD-CDMA) and multi-carrier CDMA as air interfaces, indicating that CDMA as multiple access technology will be of significant importance in future mobile communication systems. What is also of interest in the WCDMA standard, is that the use of space-time techniques and SDMA concepts is fully supported as a means to increase system capacity [36, 183, 70].

2.1 RELEVANT CELLULAR TERMINOLOGY

Cellular access systems rely on the fact that users of a single resource will be separable in one or more domain, that is frequency (FDMA), time (TDMA) or code (CDMA). Thus, in a FDMA system (for example AMPS [308] and CT-2 [47]), simultaneous transmissions to a base station will have different carrier frequencies and therefore will not interfere with one another. Similarly, in a TDMA system (for example the access method employed in one frequency allocation in an IS-54 system [47]), transmissions to the base station are separated in time to prevent interference. These multiple access techniques can also be combined to form, for example, TDMA/FDMA (IS-54) or FDMA/CDMA (IS-95 [47]) systems or any combination thereof. All of the above mentioned multiple access techniques do, however, share one common trait, the non-homogeneous geographical distribution of their subscribers. This means that all of the mentioned multiple access systems can exploit another dimension, the spatial dimension, to increase system capacity or cellular spectral efficiency.

In [122], cellular spectral efficiency is defined as a basis for rating the performance of a cellular system. Many definitions of cellular spectral efficiency have been proposed, including $bit/s/Hz$ (with the data rate measured at some predefined bit error rate (BER)), $Erlang/MHz/km^2$, equivalent telephone Erlang per square kilometer [39] and even Mbit/s-per-floor for indoor environments [39]. Because SDMA and space-time systems rely on spatial parameters, a spatial parameter is included in the definition of spectral efficiency in order to evaluate cellular system performance.

Cellular spectral efficiency (η_c) of a system is defined as the sum of the maximum data rates that can be delivered to subscribers affiliated to all base stations in a re-use cluster of cells, occupying a defined physical area. Mathematically, cellular spectral efficiency is defined as

$$\eta_c = \frac{\sum_{j=1}^{r_c} \sum_{i=1}^{K} \mathcal{R}_{ij}}{\mathcal{BW}} \frac{1}{A_{cluster}} \quad \text{bit/s/Hz/km}^2 \qquad (2.1)$$

where r_c denotes the number of cells in a re-use cluster, $\mathcal{R}_{ij}$ denotes the data rate measured in $bits/s$ at some predefined BER available to subscriber i in cell j of the re-use cluster, $\mathcal{BW}$ denotes the total signal bandwidth measured in Hz allocated to all cells in the re-use cluster, K the total number of users in a cell and $A_{cluster}$ denotes the physical area, measured in km^2, occupied by the re-use cluster. Clearly, the concept of the re-use cluster is fundamental to the determination of η_c.

Re-use cluster (r_c) A set of cells which has access to the total time/frequency and code (T/F/C) resources available in the cellular system.

Figure 2.1 shows this scenario, for the case where $r_c = 3$. Thus, each set of cells forming a re-use pattern exists totally independently (as far as T/F/C resources go) of the other cells in the area. Examining (2.1), it is clear that reducing the physical size or area of the re-use cluster will increase the cellular spectral efficiency. However, a reduced cell size will significantly increase the interference present in the cell limiting the capacity of the network in the case of a CDMA system. In this case mobiles are closer together, increasing the effects of co-channel interference - the factor determining the capacity of a cellular CDMA system. The increased amount of interference, specifically in the uplink, can be overcome using a space-time technique called high sensitivity reception (HSR) [267].

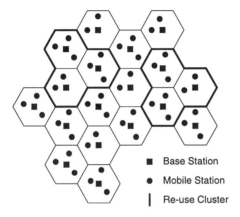

Figure 2.1.   Cellular re-use concept.

High sensitivity reception (HSR) refers to the use of adaptive antenna arrays in the uplink of a cellular network to focus the antenna beam on a specific user, thereby increasing the antenna gain in the direction of the user and suppressing transmissions received from interfering users.

This concept is depicted in Figure 2.2. In the case of TDMA and FDMA systems the HSR system may use pencil antenna beams [16] to focus on the active users, whereas in CDMA systems, the HSR system can increase the SNR in the uplink by introducing nulls in the antenna pattern in the direction of strong interfering signals.

In a manner similar to HSR, spatial filtering for interference reduction (SFIR) can be used in the downlink of a cellular system to focus all the energy radiated by the base station onto a single user or cluster of users.

Spatial filtering for interference reduction (SFIR) reduces the interference experienced by mobile communication systems in the downlink by concentrating all radiated electromagnetic energy in the direction of a user or group of users, avoiding geographical areas where no users are active.

Because the uplink of a cellular network is, in general, the capacity limiting factor, it might seem that HSR systems will yield greater capacity advantages

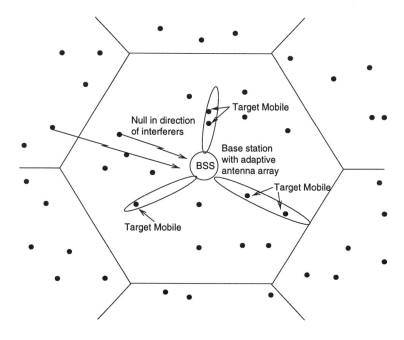

Figure 2.2. SDMA system implemented using adaptive antenna arrays.

than SFIR systems. However, the increased downlink quality afforded by SFIR techniques may lead to fewer dropped calls during hand overs (because of the better signal quality estimates available to the mobile), increasing the overall quality of service. Also, due to the dynamic nature of an adaptive antenna array, an SFIR system can facilitate the tracking of a user across cell boundaries, increasing the chances of a successful hand over to the next cell. The downlink of a cellular system in future wireless systems will also be a limiting aspect due to for example downloading of large files to a mobile terminal from a server and the limited possibilities to implement space-time signal processing techniques (multiuser detectors and space-time coding techniques) at mobile receivers. It is therefore clear that both HSR and SFIR techniques are of great importance in wireless communication systems.

Whereas HSR and SFIR techniques increase cellular spectral efficiency by decreasing the total co-channel interference levels in a cell, SDMA techniques increase cellular spectral efficiency by decreasing $A_{cluster}$. In other words, the same physical cellular network resources can be re-used more often.

Space division multiple access (SDMA) [36, 70, 183] is a multiple access technique which enables two or more subscribers, affiliated to the same base station, to use the same Time, Frequency and Code (T/F/C) resources on the grounds of their physical location or spatial separation.

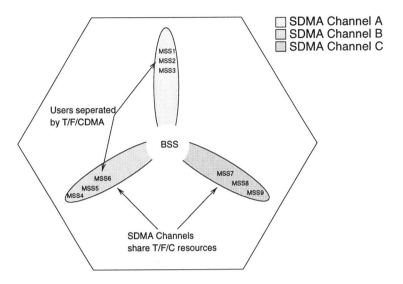

Figure 2.3. Space Division Multiple Access (SDMA) - allowing users in the same cell to share time/frequency and code resources.

This scenario is depicted in Figure 2.3, in which mobiles MSS1, MSS2 and MSS3 share the same set of T/F/C resources with MSS4, MSS5, MSS6 and MSS7, MSS8 and MSS9. For example (because of their spatial separation), MSS4 and MSS1 may both be allocated carrier frequency f_1, time slot T_1 and code a_1 although they are affiliated to the same base station. It has been shown analytically and by measurements that gains in the order of 5-9 dB can be obtained using antenna arrays with 8 elements [267].

The overall system gains that can be achieved with the techniques described above, depend heavily on the type of cellular structure employed. A plethora of different cellular structures has been defined in the literature [314, 247, 318, 96, 254, 39]. Amongst these, the best known are macro-cells, micro-cells, pico cells and hierarchical cells.

A macro-cell is a relatively large cell (2-20 km diameter) with base station antenna(s) situated well above the local urban skyline, and utilizing high levels of transmitted power (around 0.6-10 W) with rms delay spreads in the order of 0.1 to 10 μs [12].

A micro-cell is a relatively small cell (0.4-2 km diameter) with base station antenna(s) situated below the local urban skyline, and utilizing relatively low levels of transmitter power (around 20 mW) with rms delay spreads (see Chapter 3 for a detailed discussion on delay spread) in the order of 10 to 100 ns [12].

Typically, micro-cells are employed in three areas [247]

22 SPACE-TIME PROCESSING FOR CDMA

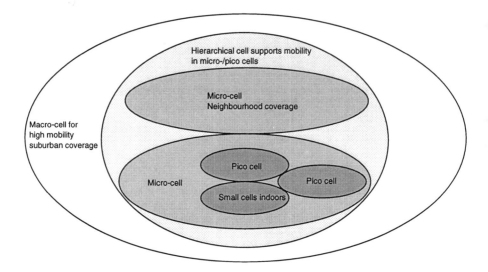

Figure 2.4. Micro-, macro-, pico and hierarchical cells work together to provide seamless service.

1. *Hot Spots* Service areas with high densities, or areas with poor coverage;
2. *Downtown Clustered Micro-cells* Urban maze of street canyons - serve pedestrians and mobiles; and
3. *In-Building* Indoor cells to serve e.g. an open plan office where users have limited mobility.

A pico cell is a small cell (less than 0.4 km in diameter) with base station antenna(s) situated at ceiling level and using very little transmitter power (0.01-0.1 W). Pico cells are typically employed inside buildings or in areas with extremely high traffic density.

An hierarchical cell covers the same geographical area as a number of micro- or pico cells, and is aimed at supplying service to subscribers with high mobility in these areas.

The relationship of the above mentioned cells is shown in Figure 2.4. In a fully operational mobile cellular environment, such as that envisioned in UMTS, all the mentioned cellular structures will be present and will work together to provide seamless service, irrespective of the subscriber's physical location or mobility [51].

Thus far, it has been shown that modern wireless access communication systems are required to supply a number of different services to users in a number of different environments (cell types). A key enabler in this regard is space-time processing techniques in which adaptive antenna arrays play a crucial role.

2.2 BASIC ADAPTIVE ARRAY CONCEPTS

An adaptive array consists of an array of spatially distributed radiating elements, with each element in the array receiving highly correlated replicas of the signal. The output of each element is adaptively weighted and combined with the other outputs to extract a specific signal from the superposition of signals received.

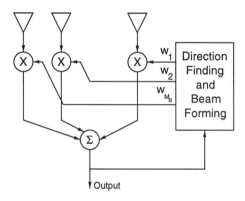

Figure 2.5. An M_B-element adaptive array with a beamforming and direction finding processor.

Figure 2.5 shows an M_B-element adaptive antenna array with a processor that performs the calculation of the direction of the desired signal, as well as the set of antenna weights required to focus the antenna radiation pattern in the direction of the desired signal. Typically, the radiating elements of an adaptive antenna array are separated by $\lambda/2$ where λ denotes the wavelength of the carrier frequency. The operation of a space-time system implemented in a TDMA environment using antenna arrays is depicted in Figure 2.2. Transmission from the target mobiles occurs at the same time instant, and the beam of the base station antenna is formed to maximize the received signal power from these target mobiles, while the received power from other interfering mobiles (inside this particular cell, as well as in adjacent cells) is minimized through the introduction of nulls in the antenna radiation pattern.

Note that in the case of systems using CDMA, the situation changes as follows. In general, it can be assumed that the number of subscribers active in a cell will be larger than the number of elements in an array, that is greater than the freedom levels of the adaptive system. All of these subscribers are transmitting in the same frequency band at the same time, meaning null steering cannot be used to cancel all interfering signals. Therefore, in the case of CDMA, at best *beamsteering* techniques in conjunction with limited null steer-

24   SPACE-TIME PROCESSING FOR CDMA

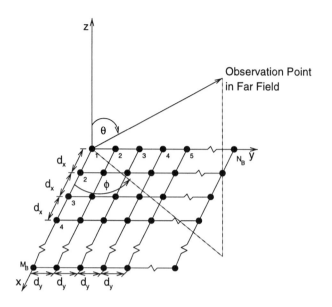

Figure 2.6. $M_B \times N_B$ element planar array.

ing can be used to point the main beam of the antenna array in the direction of a desired user or group of users [36] and to place nulls in the direction of the main interfering signals.

Many excellent texts describing the operation of antenna arrays are available, i.e. [135, 16, 11, 70], and only relevant information is summarized here. Furthermore, many different types of antenna arrays and systems are found in practice and in the literature. An incomplete list would include arrays such as uniform arrays, broadside arrays, end-fire arrays, phased (scanning) arrays, planar arrays and distributed antennas [16], which may be either fixed or adaptive. Most antenna arrays use directional antennas which have more gain in certain directions and less in others. The direction in which the gain of the antenna is maximized is referred to as the bore-sight direction of the antenna. The bore-sight gain of directional antennas is more than that of omni-directional antennas and is measured with respect to the gain of an omni-directional antenna (an omni-directional antenna has equal gain in all directions and is also known as an isotropic antenna). For example, a gain of 10 dBi (sometimes indicated by dBic or simply dB) means the power radiated by this antenna is 10 dB more than that radiated by an isotropic antenna. It should be noted that the same antenna may be used as a transmitting antenna or as a receiving antenna. In either case, the gain of the antenna remains the same. The gain of a receiving antenna indicates the amount of power it delivers to the receiver compared to an omni-directional antenna.

Some of the basic adaptive antenna arrays used for space-time applications that may be found are

uniform arrays are arrays of identical elements with identical excitation amplitudes, each with progressive phase.

broadside arrays have maximum radiation directed normal to the axis of the array ($\phi = 90°$ in Figure 2.6)

end-fire arrays have maximum radiation along the axis of the array ($\phi = 0°$ in Figure 2.6)

phased arrays are either broadside or end-fire arrays in which the phase difference between excitation currents can be controlled to point the main beam in any desired direction. The direction where the maximum gain would appear is controlled by adjusting the phase between different antennas. The phases of signals induced on various elements are adjusted so that the signals due to a source in the direction where maximum gain is required are added in-phase. This results in the gain of the array (or equivalently, the gain of the combined antenna) being equal to the sum of the gains of all individual antennas.

planar arrays are formed by placing radiating elements along a rectangular grid. These arrays are more versatile and can provide more symmetrical patterns with lower side lobes [16]. Furthermore, any other array configuration can be derived essentially from a planar array.

In practical mobile communication systems, the most frequently used adaptive antenna array is the uniform linear array (ULA) (equivalent to the array elements on the x-axis of Figure 2.6). These arrays are used because of their relative simplicity to install, and because of the practical geometry of base station masts that makes the implementation of other geometries either difficult or expensive.

The basic operation of an antenna array can be qualitatively described as follows. The total field of the array is determined by the vector addition of the fields radiated by the individual elements of the array. To create arbitrary directive patterns, it is necessary that the fields from the individual elements of the array interfere constructively in the desired directions, and destructively in the remaining directions. In an array of identical radiating elements, four main variables can be used to control the shape of the overall pattern of the antenna [16]. These are

1. the array geometry (linear, circular, etc. (see Figure 2.7));

2. the relative displacement between elements;

3. the excitation amplitude of individual elements; and

4. the excitation phase of individual elements.

Many algorithms exist for the adaptation of the weights associated with each radiating element (on a global or per-user basis) and for the combining

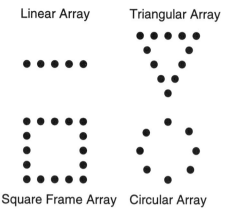

Figure 2.7. Some common array geometries.

of signals received on radiating elements. The choice of one algorithm over another is determined by various factors, such as

- rate of convergence – the number of iterations required by the algorithm to converge to the optimum solution. This aspect is of high importance in a time varying mobile channel;

- tracking – the algorithm is required to accurately track the statistical variations in the mobile environment;

- computational complexity – issues such as the number of numerical operations (i.e. multiplications, divisions etc.) and the memory size required to implement the algorithm, are of importance; and

- robustness – the algorithm should be numerical stable and should also be able to operate satisfactorily with ill-conditioned input data.

Excellent overviews of such algorithms can be found in [135, 194] with notable contributions presented, for example, in [36, 11, 131, 132, 181, 184, 272, 285]. Adaptive algorithms specifically related to CDMA will, however, be presented in Section 5.1.3.

To understand how these variables influence the design of an antenna array system, consider the $M_B \times N_B$-element planar array shown in Figure 2.6. Assuming that all radiating elements are the same (e.g. elementary dipoles), the total field of the antenna array is defined as the field of a single element, at a selected reference point (usually the origin), multiplied by the array factor of that array [16]. That is,

$$\mathbf{E}_{total}(\phi, \theta) = \mathbf{E}_{\text{single element at reference point}}(\phi, \theta) \times AF(\phi, \theta), \quad (2.2)$$

where $\mathbf{E}(\phi,\theta)$ denotes the matrix describing the electrical field of a single array element and $AF(\phi,\theta)$ denotes the array factor of the array. Alternatively, the total field can be written as

$$\mathbf{E}_{total}(\phi,\theta) = \mathbf{w}^H(\phi,\theta)\mathbf{s}_a(\phi,\theta), \quad (2.3)$$

where $\mathbf{w}^H(\phi,\theta)$ denotes the weight vector of the array, $()^H$ denotes the Hermitian transpose and $\mathbf{s}_a(\phi,\theta)$ denotes the steering vector of the array. Given a plane wave incident from direction (ϕ,θ), the steering vector describes the phase of the signal available at each antenna element relative to the phase of the signal at the reference element (usually element 1). Thus, for the planar array in Figure 2.6, the steering vector can be written as

$$\mathbf{s}_a(\phi,\theta) = \sum_{n=1}^{N_B}\sum_{m=1}^{M_B} e^{j(m-1)\mathcal{X}} e^{j(n-1)\mathcal{Y}}, \quad (2.4)$$

where $\mathcal{X} = \chi d_x \cos\phi\sin\theta + \iota_x$, $\mathcal{Y} = \chi d_y \cos\phi\sin\theta + \iota_y$, $\chi = 2\pi/\lambda$ and ι_x and ι_y denotes the progressive phase shifts between the excitation currents to elements along the x and y axes respectively. Most often, the steering vector is presented in vector or matrix form to yield

$$\mathbf{s}_a(\phi,\theta) = \begin{pmatrix} 1 & \cdots & e^{j(M_B-1)\mathcal{X}} \\ \vdots & \ddots & \vdots \\ \vdots & e^{j(m-1)\mathcal{X}} & \\ & e^{j(n-1)\mathcal{Y}} & \vdots \\ e^{j(N_B-1)\mathcal{Y}} & \cdots & e^{j(M_B-1)\mathcal{X}} \\ & & e^{j(N_B-1)\mathcal{Y}} \end{pmatrix}. \quad (2.5)$$

A set of steering vectors over all values of ϕ and θ is called the array manifold [149].

In order to control the radiation pattern of the antenna array, each of the outputs of the individual elements can be weighted by a factor w_{nm}, where w_{nm} can be any complex number. The task of a beamforming algorithm is then to find a vector $\mathbf{w}$ which, together with the steering vector, will yield a specific radiation pattern. With this in mind, it should be clear that the planar array in Figure 2.6 can be used to represent most array configurations. For example, consider the linear array depicted by the elements on the x-axis of Figure 2.6. In this case, $w_{nm} = 0$ for all $n > 1$. Thus, considering only radiation in the xy-plane ($\theta = 90°$) and assuming that excitation currents are all in phase ($\iota_x = 0$), (2.3) reduces to the well known [16, 135]

$$\mathbf{E}_{total}(\phi) = \begin{pmatrix} w_1 & w_2 e^{j\chi d_x \cos\phi} & \cdots & w_{M_B} e^{j(M_B-1)(\chi d_x \cos\phi)} \end{pmatrix}^\mathsf{T}. \quad (2.6)$$

In conclusion then, the output of the array is simply a weighted superposition of the incoming wavefronts received by each sensor or array element or

$$\mathbf{r}_o(t) = \mathbf{w}^H \mathbf{r}(t), \qquad (2.7)$$

where $\mathbf{r}(t)$ denotes the received signal as defined in (2.15).

Assuming a circular cell with a base station in the middle of the cell, Figure 2.8 depicts the radiation pattern (also called the array response, array pattern or beampattern) of a three element ($M_B = 3$) ULA with a number of uniformly distributed users in the cell. The plot shows the power received by the array at its output from a particular direction due to a unit power source in that direction. The process of combining the signals from different elements is known as beamforming, with the direction of maximum gain called the beampointing direction. From the plot it is clear that the array pattern drops to a low value on either side of the beampointing direction and the low value is called the null of the beampattern. In the ideal case the null is a position where the array response is zero, however, in practice the null is used to indicate a minimum value in the array response. The pattern between the two nulls on either side of the beampointing direction is known as the main lobe. In Figure 2.8 the main lobe(s) are in in the 90° and 180° direction. The half-power beamwidth is the width of the main lobe between the two half-power points. A smaller beamwidth results from an array with a larger extent or aperture, which is measured as the distance between the two furthest elements in the array. The gain and phase applied to the signals derived from each element may be thought of as a single complex quantity, referred to as weighting applied to the signal. If there is only one antenna element, no amount of weighting can change the pattern of that antenna. With M_B elements ($M_B > 1$), however, changing the weighting of one element relative to others may adjust the pattern to the desired value at one place, that is, one is able to place $M_B - 1$ minima or maxima anywhere in the pattern. The degrees of freedom of the antenna array is therefore $M_B - 1$. It should be noted at this stage that Figure 2.8 is an ideal plot where it is assumed that all signals arriving at the antenna array are exactly the same signals, having the same fading characteristics, but only delayed in phase due to the displacement of the antenna elements. In other words, it is assumed that the signals are completely correlated. The effect of correlation on an antenna array is investigated in Section 5.1.4.

2.3 SPACE-TIME PROCESSING

In order to simplify the analysis of space-time processing systems, a basic model of the communication system which identifies inputs, outputs and the channel is required. For a general space-time processing system where multiple antennas are employed at both the transmitter and the receiver, such a signal model is known as a multiple-input/multiple-output (MIMO) model. Clearly this is due to the fact that the desired signal has multiple inputs into the channel (the transmit antennas) as well as multiple outputs (the receive antennas). Furthermore, a MIMO system can be viewed as multiple single-input/single-

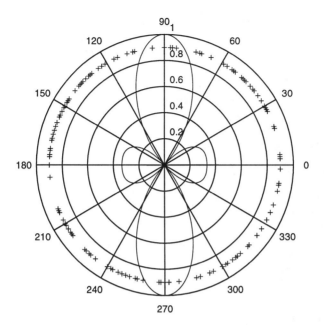

Figure 2.8. Normalized array response as a function of angle of a ULA with $M_B = 3$ and uniform user distribution ('+' indicates user).

output (SISO) sub-channels. The MIMO system's channel capacity is then the sum of the individual capacities of these sub-channels. Fading correlation effects as described in Chapters 3 and 4, affects the MIMO system capacity by modifying the distributions of the gains characterizing the SISO sub-channels. As the general MIMO case is not frequently used in practice, a number of alternative channel configurations for single user (SU) and multiuser (MU) scenarios are considered. These are

- Uplink and downlink (single-input/single-output - SISO)

  - SU-SISO: Single user with single antenna input/output at the base station and single antenna input/output at the mobile.
  - MU-SISO: Multiuser with single antenna input/output at the base station and single antenna input/output at each of the mobile units.

- Downlink (multi-input/single-output - MISO)

  - SU-MISO: Single user with multiple antenna inputs at the base station and single antenna output at the mobile unit.
  - MU-MISO: Multiuser with multiple antenna composite inputs at the base station and single antenna output at each mobile.

- Uplink (single-input/multi-output - SIMO)
    - SU-SIMO: Single user with single antenna input at the mobile and multiple antenna outputs at the base station.
    - MU-SIMO: Multiuser with single antenna input at each mobile and multiple antenna composite outputs at the base station.

As a general case then, Figures 2.9 and 2.10 depict a multiple antenna CDMA transmitter and a multiple antenna CDMA receiver respectively. As a basic receiver model applicable to both the base station and the mobile, we see that the receiver uses $M_D \times M_B$ antenna elements. Specifically, M_D diversity branches are present and each diversity branch is constituted by a M_B element beamformer. Typically, beamforming will not be used at the mobile so that $M_B = 1$ yielding a pure diversity system. Alternatively, the base station may use only beamforming ($M_D = 1$) or a combination of beamforming and diversity. At the transmitter, the user data may be coded using space-time coding techniques as described in Chapter 8, before being modulated and transmitted over M_T antennas. When the transmitter at a mobile is considered, clearly the number of data streams shown in Figure 2.9 is equal to one, whereas the number of data streams to be coded and mapped onto the transmitting antennas is equal to K, the number of users at the base station.

Turning our attention to the uplink specifically, it is assumed that the K users transmit asynchronously over a fading channel, as would be the case in any mobile system. Each user transmits by employing a different spreading sequence and the signal transmitted by the kth user is randomly delayed relative to other users. Therefore, the receiver of user k will need to recover the original data from the a composite of the desired signal, AWGN and MAI, all corrupted by a fading process. Specifically, this book looks at ways and means of using multiple antennas at both the transmitter and the receiver to make this recovery of the transmitted data more reliable.

In general, receiver structures have focused on optimizing algorithms in either the time domain or the time and code domains. Examples of such receivers include

- Time-only receivers
    - SU receivers
        * Simple correlator receiver [205]
        * RAKE receivers [205]
    - MU receivers (time and code domain)
        * Optimal (maximum likelihood) detectors [297]
        * Decorrelating receivers [300]
        * MMSE receivers [169]
- Time-only blind receivers [194]

BASIC SPACE-TIME ASPECTS 31

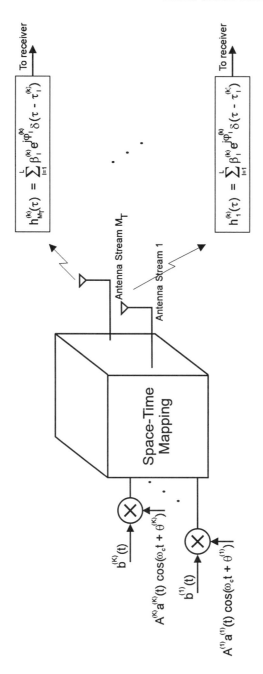

Figure 2.9. Basic block diagram of a multiple antenna CDMA transmitter.

32 SPACE-TIME PROCESSING FOR CDMA

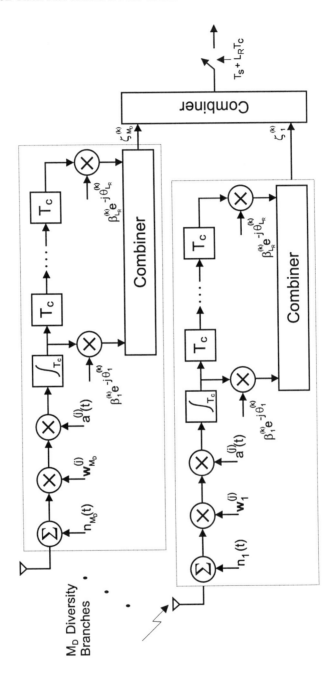

Figure 2.10. Basic block diagram of a multiple antenna CDMA receiver.

The introduction of spatial aspects into the cellular problem through the innovative use of antennas now offers new possibilities to extend the receiver algorithms mentioned above. Specifically, the use of multiple antennas at both the transmitter and the receiver adds a new dimension to the CDMA receiver problem as it allows for the improved separation of user's signals. Through the use of space-time processing techniques, the levels of MAI and fading a receiver has to cope with can be significantly reduced, thereby increasing the capacity of the overall system. In Chapters 6, 7 and 8 specifically, the above mentioned receiver structures used in conjunction with space-time processing are discussed in detail.

With this in mind, the purpose of space-time processing systems should become clearer. Essentially, space-time processing techniques provide an integrated approach to fight channel impairments on two fronts. Firstly, by introducing diversity into the system to minimize the effects of fading on the received signal and secondly by adaptively changing the radiation pattern of the antenna system to minimize the total MAI seen by the receiver. Whereas both techniques are well known, the power of space-time processing lies therein that the basic principles of beamforming and diversity are incorporated in the overall system design. Thus, space-time processing is defined as

space-time processing is the minimization of fading and MAI through the integrated use of multiple antennas, advanced signal processing techniques, advanced receiver structures and forward error correction.

From this definition it is clear that techniques such as SFIR, HSR, SDMA and various others are all different forms of space-time processing where the emphasis of the space-time processing technique is on different ways to reduce fading and MAI. Furthermore, smart antennas can be then be defined as

smart antennas is the combination of antennas with signal processing algorithms to yield an antenna system with dynamic properties.

These dynamic properties may, for example, be a radiation pattern that changes according to the motion of the mobile.

Based on these definitions, the main aim of space-time techniques for mobile systems is to maintain an acceptable level of error performance and, hence, to maximize the signal-to-interference and noise ratio (SINR) for each user in the system. An antenna array containing M_B-elements can provide a mean power gain of M_B over white noise, but suppression of interference from other cellular users is dependent on the form of the received data. Two cases are of interest to a cellular CDMA system.

The first is when the processing gain, N, of the CDMA system is small, so that the interference contribution from another user on the same channel will be significant. An antenna array can null-out up to $(M_B - 1)$ users on the uplink [166], so significant performance improvements may be possible. Secondly, if the processing gain N is large, a good model for the received signal in a power controlled CDMA system is a strong desired signal corrupted by a large number of small cross-correlation terms. In this case nulling out $(M_B - 1)$ interferers is unlikely to significantly improve the received SINR [322] because of the very large number of interference components. An improved method here would be to make full use of the space-time processing power, and to estimate the form of the received signal and determine the multiuser solution or to add additional error control coding [179]. These last two techniques are the subjects of Chapters 7 and 8 respectively. This type of receiver can exploit any spatial diversity present, while suppressing the mean level of CDMA interference by a factor proportional to M_B.

2.3.1 Space-time Base Stations

A base station with space-time techniques can take on many different forms [174, 261, 11]. In its simplest form, multiple antennas at the base station may be used to form multiple fixed beams to cover the whole cell site. For example, three beams with a beamwidth of 120° each or six beams with a beamwidth of 60° each may be used. Each beam may then be treated as a separate cell, and the frequency assignment may be performed in the usual manner, with mobiles handed to the next beam as they leave the area covered by the current beam, using the normal hand-off process when a mobile crosses a cell boundary.

Alternatively, at the base station the antenna array may be used to find the location of each mobile, and then beams can be formed to cover different mobiles or groups of mobiles. Each beam can be considered as a co-channel cell, and thus may be able to re-use a system frequency or code. This setup is different from the fixed beam case, where a number of beams of fixed shape cover the whole cell. Here, the beams are shaped to cover the traffic. As the mobiles move, the different beams cover different clusters of mobiles, offering the benefit of transmitting the energy toward the mobiles. This arrangement is particularly useful in situations where the mobiles move in clusters or along confined paths, such as highways. It is envisaged in [261] that each mobile may be covered by a separate beam, reducing the hand-off problem. The concept of adaptive beamforming may be extended to dynamically changing cell shapes. Instead of having cells of fixed size, the use of array antennas allows the formation of a cell based upon traffic needs [174]. To realize such a base station, the ability to locate and track mobiles would be required in order to adapt the system parameters to meet traffic requirements.

Another important technology related to dynamic or adaptive beamforming is so-called Intelligent technology. This technology allows cell shapes and sizes to be changed based upon traffic conditions, channels to be assigned dynamically as per traffic needs, and transmitter power to be adapted according to receiver requirements [174]. Apart from variable cell size, adaptive transmitted power, and dynamic channel allocation, this technology also allows for variable speed of transmission and the use of adaptive modulation and demodulation techniques [174, 186, 270, 328, 327].

In contrast to steering beams toward the desired mobile, one may adjust the antenna pattern so that it has nulls toward other mobiles, thereby reducing interference. Ideally an antenna pattern with zero response in the nulls should be generated. However, this is seldom achievable and a pattern is usually created with a reduced response toward undesirable interferences. Formation of nulls in the antenna pattern toward co-channel mobiles helps to reduce the co-channel interference in two ways. In transmit mode, less energy is transmitted from the base toward these mobiles, thereby reducing the interference from the base station. In receive mode, this helps to reduce the interference contribution from these mobiles.

It is usually assumed that the base station knows where the desired (or undesired) mobile is so as to either adaptively track the desired mobile (or group of mobiles) or adaptively nullify the undesired mobile(s). For the base station to do this, it is often necessary for it to acquire knowledge of the direction of arrival (DOA) of users. Many of the DOA estimation techniques available in the literature assume the existence of point sources. This is hardly ever the case for land-mobile communications, where scattering in the vicinity of a mobile causes a spreading of the source and the signal arrives at the base in a multi-path form, as if there were many radiating sources of varied power in the neighborhood of the mobile. This is strongly related to the environment in which the system will operate and will be considered in detail in Chapter 4. To estimate the DOA or steering vector, classical methods such as the MUSIC [239] and ESPRIT [195] methods can be used. However, in a CDMA system, such as the UMTS system, the number of users will far exceed the number of antenna elements. Therefore, subspace methods such as the ones mentioned are not always applicable. In Chapter 5 methods to perform optimal spatial filtering using adaptive techniques and methods to estimate the DOA will be discussed. A reference signal can also be used to determine the position of the mobile. Initial weights can be estimated by using a specific signal transmitted for this purpose. Once the initial weights are determined, the received signal may serve as the reference for later updating the weights and tracking the mobile. In a CDMA system, a separate sequence may be transmitted for this purpose, or the system synchronization sequence may be used.

2.3.2 Space-time Mobiles

Due to space limitations on current mobile phones, implementation of space-time techniques is not yet a reality. It is expected that this will change in future as new antenna structures and more power-efficient digital signal processors become available. Developments in dual antenna systems for hand-held portables are reported on in [19, 277, 54]. However, not only mobile phones will be available in the near future. Other types of mobile terminals, e.g. notebook type terminals, will incorporate space-time processing techniques.

2.3.3 Advantages of Space-time Processing

As has been discussed in [322, 20, 174, 89, 19, 261, 90], space-time techniques have the ability to improve the performance of a mobile communication system in a number of ways. Specific advantages of space-time techniques are that they yield:

- Increased capacity (spectrum efficiency) by increasing the number of active users for a given BER quality.

- Reduction of co-channel interference to improve service quality and/or increase the frequency re-use factor. This point is especially important in CDMA-based systems in which the system capacity is interference limited.

- Reduction in delay spread and fading. By beamforming and diversity techniques, the SINR of the system can be improved in a fading environment. Related to this is the reduction of the effect of angular spreading of the received signal due to scatterers around the mobile (which is close to the ground) by narrow beams being formed on the arriving signals.

- Reduction in outage probability. Outage probability is the probability of a channel being inoperative due to an increased error rate. For example, by reducing interference using space-time techniques, the outage probability can be reduced.

- Increase in transmission efficiency. Due to the high directivity and gain of the space-time system, base station range may be extended, and a mobile may be able to transmit using less power resulting in longer battery life.

- Dynamic channel assignment. Making use of SDMA, channels may be dynamically assigned as a function of the traffic demand in the cell.

- Reduction in hand-off rate. When the capacity of a cellular system is exceeded, cell splitting is used to create new cells, each with its own base station and new frequency assignment, with increased hand-off as a result. This may be reduced by space-time processors which can create independent beams.

- Improved positioning accuracy by applying antenna arrays.

- Reduction in cost, complexity and potential network architecture simplification. There is no doubt that dynamic hand-off, dynamic channel assignment, dynamic and dynamic nulling (all features of space-time systems) require more complexity. However, careful consideration should also be given to the overall improvement of system reliability, quality of service, etc. when comparing space-time systems to conventional systems.

While antenna arrays provide many advantages, these must be offset against cost and complexity factors. A number of important points to be considered are

- hardware and software requirements increase as the number of antenna elements increases; and

- in practical situations, the antenna array performance may be adversely affected by channel modeling errors, calibration errors, phase drift and noise which is correlated between antennas.

2.4 SYSTEM MODEL FOR SPACE-TIME PROCESSING

With reference to Figures 2.9 (for the case where $M_T = 1$) and 2.10, the output for user k can be written as

$$s^{(k)}(t) = \text{Re}\left\{A^{(k)} b^{(k)}(t) a^{(k)}(t) \exp\left[-j\left(\omega_c t + \theta^{(k)}\right)\right]\right\}, \tag{2.8}$$

where $A^{(k)}$ denotes the received signal amplitude[1], $b^{(k)}(t)$ denotes a binary data sequence with symbol period T_s seconds, i.e.,

$$b^{(k)}(t) = \sum_{i=0}^{P} b_i^{(k)} u(t/T_s - i), \tag{2.9}$$

where

$$u(t) = \begin{cases} 1 & 0 \leq t \leq 1 \\ 0 & \text{otherwise} \end{cases}, \tag{2.10}$$

with $P+1$ symbols transmitted, $a^{(k)}(t)$ denotes a binary spreading waveform with a processing gain of N, leading to chips of duration $T_c = T_s/N$. If the period of the spreading waveform is equal to T_s, then the equivalent spreading sequences are denoted short codes as opposed to the case when the period is far greater than T_s where these sequences are termed long codes. Also, standard binary phase shift keying (BPSK) modulation is used with a carrier frequency of ω_c rad/s and unknown carrier phase θ_k, a random variable uniformly distributed over $[0, 2\pi)$. The transmitted signal is propagated over a radio channel modeled as a time-varying, discrete multi-path fading channel with equivalent low-pass response

$$h_i^{(k)}(\tau) = \sum_{l=1}^{L} \beta_l^{(k)}(i) \cdot \exp\left(j\varphi_l^{(k)}(i)\right) \cdot \delta[\tau - \tau_l^{(k)}]. \tag{2.11}$$

Each path is characterized by the random variables $\beta_l^{(k)}(i)$ denoting the strength of path l from user k at symbol interval i. Each of these can be modeled as either Rayleigh distributed or Nakagami-m distributed, depending on the overall channel model. Each path is also associated with the phase shift parameter $\varphi_l^{(k)}(i)$, uniformly distributed over $[0, 2\pi)$ and the propagation delay $\tau_l^{(k)}$, uniformly distributed over $[0, T_s)$. We assume here that multi-path delays do not change with time. Define the complex fading process in time for user k, multi-path component l as

$$\hat{c}_l^{(k)}(t) = \sum_{i=0}^{P} c_l^{(k)}(i) u(t/T_s - i), \qquad (2.12)$$

where

$$c_l^{(k)}(i) = \beta_l^{(k)}(i) \exp\left(j\varphi_l^{(k)}(i)\right). \qquad (2.13)$$

The time variations of the fading for each path are governed by the corresponding Doppler spectrum, while the multi-path time delays are assumed to remain constant. Assuming that L multi-path components are present and that reception is done by an M_B element ULA, the received signal for element m_B can be written as

$$\begin{aligned} r_{m_B}(t) &= \sum_{k=1}^{K}\sum_{l=1}^{L} A^{(k)} b^{(k)}(t-\tau_l^{(k)}) a^{(k)}(t-\tau_l^{(k)}) \hat{c}_l^{(k)}(t-\tau_l^{(k)}) \qquad (2.14)\\ &\quad \times \exp\left[-j\left(\omega_c t + \theta^{(k)}\right)\right] s_{a_{m_B}}(\phi_l^{(k)}) + n_{m_B}(t), \end{aligned}$$

where $\phi_l^{(k)}$ is the broadside DOA for multi-path component l of user k, $s_{a_{m_B}}$ is the m_Bth component of the steering vector of diversity branch M_D, and $n_{m_B}(t)$ denotes additive complex white Gaussian noise with a two-sided spectral density of N_0. It is assumed that the angles of arrival stay constant over the detection interval.

Since coherent demodulation is assumed, the receiver is provided with complete knowledge of the carrier phase, multi-path signal phases and multi-path time delays. Defining

$$\mathbf{r}(t) = (r_1(t), r_2(t), \cdots, r_{M_B}(t))^\mathsf{T}, \qquad (2.15)$$

and

$$\mathbf{n}(t) = (n_1(t), n_2(t), \cdots, n_{M_B}(t))^\mathsf{T}. \qquad (2.16)$$

Assuming that the desired signal is that of user j, the output of the receiver for symbol interval i after correlation, demodulation, RAKE combining and sampling can be written as

$$\begin{aligned} \zeta_{m_D}^{(j)}(i) &= \sum_{n=1}^{L_R} \int_{iT_s+\tau_q^{(j)}}^{(i+1)T_s+\tau_q^{(j)}} \exp\left[-j\left(\omega_c t + \theta^{(j)}\right)\right] a_k(t-\tau_n^{(j)}) \frac{\left(\mathbf{w}^{(j)}\right)^\mathsf{H}}{\|\mathbf{w}^{(j)}\|} \mathbf{r}(t) dt \\ &= \sum_{n=1}^{L_R} S_n^{(j)}(i) + I_{sin}^{(j)}(i) + I_{main}^{(j)}(i) + \eta_n^{(j)}(i), \qquad (2.17) \end{aligned}$$

where

$$S_n^{(j)}(i) = \|\mathbf{w}^{(j)}\|\overline{\mathcal{R}}^{(jj)} A_j \frac{T_s}{2} b_i^{(j)} c_n^{(j)}(i), \tag{2.18}$$

$$I_{sin}^{(j)}(i) = \sum_{\substack{l=1 \\ l \neq n}}^{L} \frac{A_j}{2} c_l^{(j)}(i) \|\mathbf{w}^{(j)}\|\overline{\mathcal{R}}^{(jj)} \tag{2.19}$$

$$\times \left\{ b_{i+1}^{(j)} R_{jn,jl}^{(1)}(i) + b_i^{(j)} R_{jn,jl}^{(0)}(i) + b_{i-1}^{(j)} R_{jn,jl}^{(-1)}(i) \right\},$$

$$I_{main}^{(j)}(i) = \sum_{\substack{k=1 \\ k \neq j}}^{K} \sum_{l=1}^{L} \frac{A_k}{2} c_l^{(k)}(i) \exp\left[-j\left(\theta^{(j)} - \theta^{(k)}\right)\right] \|\mathbf{w}_l^{(k)}\|\overline{\mathcal{R}}^{(jk)} \tag{2.20}$$

$$\times \left\{ b_{i+1}^{(k)} R_{jn,kl}^{(1)}(i) + b_i^{(k)} R_{jn,kl}^{(0)}(i) + b_{i-1}^{(k)} R_{jn,kl}^{(-1)}(i) \right\},$$

and

$$\overline{\mathcal{R}}^{(jk)} = \frac{\left(\mathbf{w}^{(j)}\right)^{\mathrm{H}} \mathbf{w}^{(k)}}{\|\mathbf{w}^{(j)}\|\|\mathbf{w}^{(k)}\|}, \tag{2.21}$$

with $(\cdot)^{\mathrm{H}}$ denoting the Hermitian transpose and $\mathbf{w}^{(j)}$ the weight vector optimizing the SINR of user j or

$$\mathbf{w}^{(j)} = (w_1, w_2, \cdots, w_{M_B})^{\mathrm{T}}. \tag{2.22}$$

Furthermore, $b_i^{(j)}$ denotes the information bit for symbol interval i of user j, and also

$$R_{jn,kl}^{(-1)}(i) = \begin{cases} \int_{iT_s+\tau_n^{(j)}}^{iT_s+\tau_n^{(j)}+\tau_l^{(k)}} a_k(t-\tau_l^{(k)})a_j(t-\tau_n^{(j)})dt; & \tau_l^{(k)} - \tau_n^{(j)} > 0 \\ 0 & \text{otherwise,} \end{cases}$$

$$\tag{2.23}$$

$$R_{jn,kl}^{(0)}(i) = \begin{cases} \int_{iT_s+\tau_n^{(j)}}^{iT_s+\tau_n^{(j)}+\tau_l^{(k)}} a_k(t-\tau_l^{(k)})a_j(t-\tau_n^{(j)})dt; & \tau_l^{(k)} - \tau_n^{(j)} < 0 \\ \int_{iT_s+\tau_n^{(j)}+\tau_l^{(k)}}^{(i+1)T_s+\tau_n^{(j)}} a_k(t-\tau_l^{(k)})a_j(t-\tau_n^{(j)})dt; & \tau_l^{(k)} - \tau_n^{(j)} > 0, \end{cases}$$

$$\tag{2.24}$$

$$R_{jn,kl}^{(1)}(i) = \begin{cases} \int_{iT_s+\tau_n^{(j)}+\tau_l^{(k)}}^{(i+1)T_s+\tau_n^{(j)}} a_k(t-\tau_l^{(k)})a_j(t-\tau_n^{(j)})dt; & \tau_l^{(k)} - \tau_n^{(j)} < 0 \\ 0 & \text{otherwise.} \end{cases}$$

$$\tag{2.25}$$

The noise sample $\eta_n^{(j)}(i)$ is a zero-mean complex Gaussian distributed random variable with variance $\sigma^2 = N_0 T_s/2$.

2.5 PRACTICAL IMPLEMENTATION OF CDMA SMART ANTENNA SYSTEMS

The real power of smart antenna systems lie in the use of advanced software radio architectures to implement adaptive beamforming and diversity combining schemes. Specifically, the software radio allows one to build flexible, multi-band radio systems. The aim of the software radio principle is to have a general hardware platform that can principally cope with the strongest constraints of all the mobile communications standards or configurations to be supported by the platform [105]. Such a radio may be dynamically updated with new software without changes in the hardware or infrastructure. This software may either have an IF software radio architecture, in which the waveforms are synthesized and received in software, or they may have a baseband DSP radio architecture, in which the software is limited to bit stream signal processing. In general, the IF based architecture will require substantially more processing power than the baseband architecture.

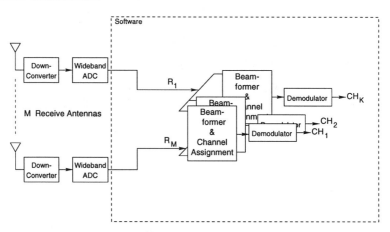

Figure 2.11. Functional block diagram of a software radio for a base station with smart antenna.

A functional block diagram of a typical software radio based smart antenna system is shown in Figure 2.11 [218]. As illustrated in this figure, a wideband front end downconverts the received signal to an IF where it is sampled and digitized by a high speed ADC. The rest of the processing is implemented in software. In Figure 2.11, each of the M antenna elements has its own downconverter and ADC with the subsequent beamforming and demodulation implemented in software and shared among all of the elements. This beamforming process can be implemented using DSPs or in hardware [127, 33]. Usually, adaptive algorithms are used to update the weight vectors. The weight vectors used are typically computed adaptively, with or without the use of training sequences (blind detection). To this end a variety of adaptive algorithms may be used, such as the least mean-square (LMS) algorithm, least square lattice (LSL) algorithm and the fast transversal filter (FTF) algorithm.

The power of combining smart antenna systems and software radio algorithms can be described using an example of power control and beamforming. In [218] it is shown that distributed power control algorithms converge to the optimal power allocation that minimizes the transmitted power for each mobile while the link quality is guaranteed for each receiver.

Joint power control and beamforming algorithms are proposed in [217, 211, 212]. In these algorithms, the beamforming weight vectors and power allocations are updated jointly by a distributed algorithm. The weight vectors are updated by the beamforming algorithm only during a training phase, and are constant between the training intervals. The complexity of the first step of the power control algorithm is $4M + 1$ multiplications and one addition for the calculation of the SINR and one division and addition for updating the power. Typical simulation results comparing the power control algorithms with and without antenna arrays are shown in Figure 2.12.

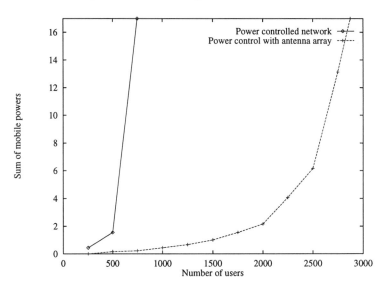

Figure 2.12. The total mobile powers versus the number of users.

2.6 SUMMARY

This chapter forms the first of three chapters that describes basic space-time concepts. In Chapters 3 and 4 the various aspects influencing space-time performance and correlation effects will be discussed respectively. Specifically, this chapter has summarized some of the basic principles associated with space-time processing. Cellular terminology relevant to space-time processing, such as SFIR, HSR and SDMA were defined. Adaptive antenna arrays were discussed as one of the key technologies in space-time processing and a general space-time system model, which will form the basis of later discussions on space-time processing, has been presented.

Notes

1. Slow power control is assumed such that $A^{(k)}$ is adjusted to compensate for path loss and shadowing.

3 ASPECTS INFLUENCING SPACE-TIME PERFORMANCE

The gain offered by space-time processing relies on many parameters, some of which are beyond the control of the design engineer but which should be modeled accurately when analyzing such systems. Specifically four main areas of influence can be identified namely: (i) the propagation path of the signal, (ii) temporal fading, (iii) the scattering environment and, (iv) the angular distribution of subscribers. These factors influence the system performance and careful attention should be given to the different aspects for optimal system design. For instance, it has been shown by Marzetta *et al.* [170] that there is no advantage in making the number of transmit antenna M_T greater than the length of the channel coherence interval (described in Section 3.2.4). A thorough understanding of the various aspects influencing space-time performance is therefore needed.

3.1 PROPAGATION PATH

The modeling of the propagation path needs to take into account a number of effects. These include

- Path (propagation) loss;

- Shadowing (The particular scattering environment (i.e. trees, buildings) along a path at a given distance will be different for every path, causing variations with respect to the nominal value given by the path loss model.

Some paths will suffer increased loss, while others will be less obstructed and have an increased signal strength. This phenomenon is called shadowing or slow fading and has log-normal fading statistics.);

- Number of multi-path components and the distribution of their envelopes (These effects are a result of the local scattering environment around the mobile and/or base station);

- Temporal fading (Due to its fundamental importance in a mobile environment, this effect is described in detail in Section 3.2, with emphasis on a space-time fading environment); and

- Correlation (Multi-path components generated by a single area of local scatterers may show considerable correlation, with the correlation depending heavily on assumptions made concerning the spatial distribution of local scattering elements. Correlation is a very important concept in space-time systems since it influences the antenna pattern in beamforming (see Section 5.1.4) and the amount of diversity gain achievable in the system. The effect of correlation is considered in detail in Section 4.2.2.1, after the discussion of appropriate channel models for space-time systems).

The above mentioned propagation characteristics influence mainly the performance of the beamforming algorithm used, as well as the performance of the combining algorithm used in the case of space-time systems relying on both beam steering and diversity techniques. Most beamforming algorithms used assume that the signals arriving at each element of the array are highly correlated (ρ_{ij} >0.8) [175]. This assumption depends heavily on the composition of the local scattering area surrounding the mobile.

3.1.1 Path Loss

If a wireless channel's propagating characteristics are not specified, it is usually inferred that the signal attenuation versus distance behaves as if propagation takes place over ideal free-space. The model of free space treats the region between the transmitting and receiving antennas as being free of all objects that might absorb or reflect RF energy. It is further assumed that, within this region, the atmosphere behaves as a perfectly uniform and non-absorbing medium. Furthermore, the earth is treated as being infinitely far away from the propagating signal. In this idealized free-space model, the attenuation of RF energy behaves according to an inverse-square law. The received power expressed in terms of transmitted power is attenuated by a factor $L_s(R)$, where this factor is called the path loss or free space loss. In a mobile wireless application, the mean path loss, $\overline{L}_s(R)$, as a function of distance R between the mobile and base-station, is proportional to an n_lth power of R relative to a reference distance r_0. Therefore

$$\overline{L}_s(R) = \left(\frac{R}{r_0}\right)^{n_l}. \tag{3.1}$$

In the presence of a very strong guided wave phenomenon, such as urban streets, n_l can be lower than 2. When obstructions are present, n_l is larger, with typical values in the order of 2.5 to 5. Measurements have shown that for any value of R, the total path loss $L_x(R)$ is a random variable having a log-normal distribution about the mean distant-dependent value $\overline{L}_s(R)$ [48]. Thus, total path loss $L_x(R)$ can be expressed in terms of $\overline{L}_s(R)$ plus a random variable X_σ, as follows (stated in decibels) [209]

$$L_x(R) = L_s(r_0) + 10 \cdot n_l \cdot \log_{10}(R/r_0) + X_\sigma, \tag{3.2}$$

where X_σ denotes a zero-mean Gaussian random variable (in decibels) and is site and distance dependent. The choice of a value for X_σ is often based on measurements, and is dependent on the type of cellular environment (macro-, micro- or pico cell) and other channel parameters. Typical values are as high as 6-10 dB.

There are three basic mechanisms that impact on signal propagation in a mobile communication system [209, 250].

- **Reflection** occurs when an electromagnetic wave impinges on a smooth surface with very large dimensions compared to the RF signal's wavelength (λ).

- **Diffraction** occurs when the radio path between the transmitter and receiver is obstructed by a dense body with large dimensions compared to λ causing secondary waves to be formed behind the obstructing body. (It is a phenomenon that accounts for RF energy traveling from transmitter to receiver without a LOS path between the two. It is often termed shadowing because the diffracted field can reach the receiver even when shadowed by an impenetrable obstruction.)

- **Scattering** occurs when a radio wave impinges on either a large rough surface or any surface whose dimensions are in the order of λ or less, causing the reflected energy to spread out (scattering in all directions). In an urban environment, typical signal obstructions that yield scattering are lamp posts, street signs and foliage.

3.2 TEMPORAL FADING

Based on experimental evidence, the cause of fading can be attributed to large-scale fading and/or small-scale fading. Large-scale fading (or shadowing) has path loss as a result with effects as described in Section 3.1. Small-scale fading manifests itself in two mechanisms, namely signal dispersion (time-spreading of the signal) and time-variant behavior of the channel. Due to motion between the transmitter and the receiver the channel is time-variant as a result of the propagation path changing. The rate of change of these propagation conditions accounts for the rapidness of the fading (rate of change of the fading impairments). Small-scale fading is generally statistically described by either a Rayleigh [250, 205], Rician [205, 255] or Nakagami-m [336, 255, 178] distribution. The model choice depends mainly on the operating environment of the communication system. If the multiple reflective paths are large in number and there is no LOS signal component, the envelope of the received signal is traditionally statistically described by a Rayleigh probability density function. When there is a dominant non-fading signal component present, such as a LOS propagation path, the small-scale fading envelope is described by a Rician pdf. In addition to the attractive mathematical properties of the Nakagami-m fading model, it has also been shown in [35, 258] that the Nakagami model can be used to accurately describe the fading behavior of multi-path signals and the varying physical scattering processes.

In the following sections a number of useful power spectral density functions and correlation functions, that define the characteristics of a fading multi-path channel, will be considered. These functions are fundamental to the understanding of space-time processing techniques.

3.2.1 Multi-path Intensity Profile

Knowledge of the multi-path intensity profile (MIP), $S_M(\tau, \phi_r)$ gives an indication of how the average received power varies as a function of the time delay τ and the angle ϕ_r, the angle of arrival measured relative to the LOS from the base station and the mobile.

The term time delay is used to refer to excess delay in the time domain. It represents the signal's propagation delay that exceeds the delay of the first signal's arrival at the receiver. In a space-time system, due to the directivity of the antenna, $S_M(\tau, \phi_r)$ is also influenced by the angle of the received multi-path signals. For a typical wireless space-time channel, the received signal usually consists of several discrete multi-path components, arriving from different angles. The distribution of these multi-path components is closely related to the environment (i.e. urban, rural, etc.), cell architecture (i.e. macro-, micro-, or pico cell) and the DOA of arriving signals at the receiver. The parameter τ_m represents the maximum excess delay during which the multi-path signal power falls to some threshold level below that of the strongest component. The threshold level might be chosen at 10 or 20 dB below the level of the strongest component.

In classical wireless channels, where it is assumed that an omni-directional antenna is employed at the receiver and that the DOA of multi-path signals is uniformly distributed at the receiver, $S_M(\tau, \phi_r)$ is independent of the DOA angle ϕ_r. When a space-time system is considered, some assumption has to be made about the environment in which the system operates. In Chapter 4 various channel models for space-time systems will be considered, from which

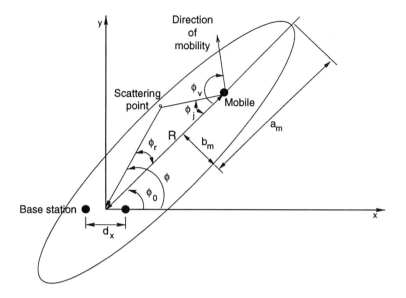

Figure 3.1. GBSBEM scatterer density geometry.

it would be possible to determine $S_M(\tau, \phi_r)$ as a function of the DOA of the multi-path components. A very important parameter in the channel model for space-time systems is the scattering environment. To illustrate the influence of the scattering environment on the MIP, consider Figure 3.1. In this model it is assumed that the scatterers are uniformly distributed within an ellipse [64], where the base station and the mobile are the foci of the ellipse. This model, known as the Geometrically Based Single Bounce Elliptical Model (GBSBEM), was proposed for micro-cell environments where the antenna heights are relatively low and multi-path scattering near the base station is just as likely as multi-path scattering near the mobile. The MIP as a function of the DOA and time of arrival (TOA) at the base station, using the GBSBEM model, is given by

$$S_M(\tau, \phi_r) = \begin{cases} \frac{(R^2 - \tau^2 c^2)(R^2 c + \tau^2 c^3 - 2\tau c^2 R \cos \phi_r)}{4\pi a_m b_m (R \cos \phi_r - \tau c)^3} & \forall \quad \frac{R}{c} \leq \tau \leq \tau_m \\ 0 & \text{otherwise} \end{cases}, \quad (3.3)$$

where

$$a_m = \frac{c\tau_m}{2}, \quad b_m = \frac{1}{2}\sqrt{c^2 \tau_m^2 - R^2},$$

c is the speed of light and R is the distance between the base station and mobile.

48 SPACE-TIME PROCESSING FOR CDMA

Using (3.3), Figure 3.2 indicates $S_M(\tau, \phi_r)$ as a function of the DOA for the GBSBEM model at the base station for $R = 1$ km and $\tau_m = 5\mu s$. From this figure it is clear that the model predicts a relatively high probability of multi-path components with small excess delays along the LOS. Therefore, from the base station's perspective, all the multi-path components are restricted to a small range of angles. Also important to note is that, for an ideal system (zero excess delay), the function $S_M(\tau, \phi_r)$ would consist of an impulse with weight equal to the total average received signal power. An important gain from space-time techniques is, therefore, to reduce the delay spread of the channel. Simulation results [162], confirmed by measurements in Toronto [252], show that a 60° beam-width antenna reduces the mean rms delay spread by about 30-34%.

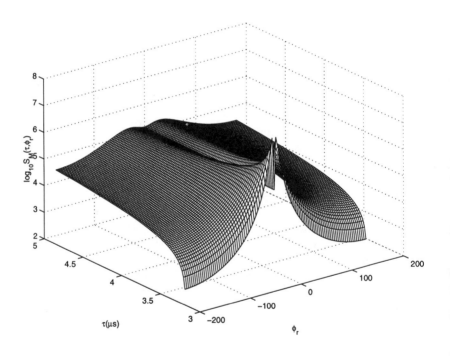

Figure 3.2. $S_M(\tau, \phi_r)$ as a function of DOA and TOA at base station for the GBSBEM model.

The relationship between maximum excess delay time, τ_m, and symbol time, T_s, can be viewed in terms of two different degradation categories, frequency-selective fading and frequency non-selective or flat fading, which will be described later.

3.2.2 Space-frequency Correlation Function

The space-frequency correlation function, $|R_M(\Delta f, \phi_r)|$, represents the correlation between the channel's response to two signals as a function of the frequency difference between the two signals. It can be thought of as the channel's frequency transfer function. Therefore, the time-space manifestation can be viewed as if it were the result of a filtering process. Knowledge of $|R_M(\Delta f, \phi_r)|$ gives an indication of the correlation between received signals that are spaced in frequency by $\Delta f = f_2 - f_1$. The coherence bandwidth, $\mathcal{BW}_{ch}$, is a statistical measure of the range of frequencies over which the channel passes all spectral components with approximately equal gain and linear phase. Thus, the coherence bandwidth represents a frequency range over which frequency components have a strong potential for amplitude correlation. That is, a signal's spectral components in that range are affected by the channel in a similar manner as, for example, in exhibiting fading or no fading. Note that $\mathcal{BW}_{ch}$ and τ_m are reciprocally related (within a multiplicative constant). As an approximation, it is possible to say that

$$\mathcal{BW}_{ch} \approx 1/\tau_m. \tag{3.4}$$

The maximum excess delay, τ_m, is not necessarily the best indicator of how any given system will perform on a channel because different channels with the same value of τ_m can exhibit very different profiles of signal intensity over the delay span. A more useful measurement of delay spread is most often characterized in terms of the rms delay spread σ_τ, where

$$\sigma_\tau = \sqrt{\overline{\tau^2}_m - (\overline{\tau}_m)^2}, \tag{3.5}$$

and $\overline{\tau}_m$ is the mean excess delay. An exact relationship between coherence bandwidth and delay spread does not exist and must be derived from actual signal dispersion measurements in particular channels using Fourier or other signal processing techniques.

To investigate the coherence bandwidth in a space-time environment, knowledge of the joint distribution of the DOA angle, ϕ_r, and σ_τ provides an almost complete description of the mobile radio channel. Thus, integration of the joint distribution with respect to ϕ_r yields the distribution of time delays. Also, the cross-covariance of two signals of different frequencies, one shifted in time by τ from the other, depends on the joint distribution of the DOA angle, ϕ_r, and the TOA, τ. The Fourier transform of this cross-covariance yields the cross-spectrum of the two frequency separated signals. Therefore, to calculate $|R_M(\Delta f, \phi_r)|$, we simply calculate the Fourier transform of $S_M(\tau, \phi_r)$ for the time delays τ. We can then write

$$|R_M(\Delta f, \phi_r)| = \left| \int_{-\infty}^{\infty} \exp(-j\Delta f \tau) S_M(\tau, \phi_r) d\tau \right|. \tag{3.6}$$

Using (3.3) and (3.6), Figure 3.3 depicts $|R_M(\Delta_f, \phi_r)|$ under similar conditions as Figure 3.2. Examining the figure for $\phi_r = 0$ or along the LOS direction, it is evident that the correlation between two signals with large frequency difference is substantial. As the signals arrive at angles away from the LOS direction, the correlation quickly reduces. Also, for small frequency offset, the correlation of signals are high regardless of the angle of arrival.

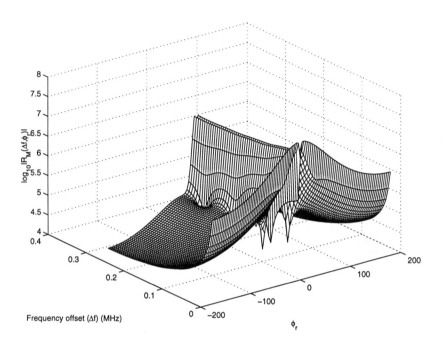

Figure 3.3. $|R_M(\Delta f, \phi_r)|$ as a function of DOA and TOA at base station for the GBS-BEM model.

Due to the relationship between τ_m and $\mathcal{BW}_{ch}$, a frequency domain definition of frequency selective and flat fading can also be derived. This is known as the Doppler spectrum.

3.2.3 Doppler Power Spectrum

To date, we have described signal dispersion and coherence bandwidth, parameters that define the channel's time-spreading properties in a local area. However, this does not offer information about the time-varying nature of the channel caused by relative motion between the transmitter and receiver, or motion of objects within the channel. For mobile applications, the channel is time-variant because motion between the transmitter and receiver results in propagation path changes. Thus, as a result of such motion, the receiver sees variations in signal amplitude and phase. Assuming that all scatterers making up the channel are stationary, whenever motion ceases, the amplitude and phase of the received signal remain constant, in other words, the channel appears to be time-invariant. Whenever motion begins again, the channel once again appears time-variant. Since the channel characteristics are dependent on the position of the transmitter and receiver, time variance in this case is equivalent to spatial variance.

The received signal at a mobile experiences Doppler spread due to the motion of the receiver. In a more practical sense, knowledge of the Doppler power spectral density, $S_D(f, \phi_r)$, indicates how much spectral broadening is imposed on the signal as a function of the rate of change of the channel state. The jth multi-path component arrives at the mobile at an angle of ϕ_j with respect to the LOS component, as indicated in Figure 3.1. The multi-path components at the receiver experience Doppler shift depending on the direction of motion of the mobile and the Doppler shift of each multi-path component is generally different from that of other paths. The effect on the received signal is seen as a Doppler spreading or spectral broadening of the transmitted signal frequency, rather than a shift. The jth multi-path component experiences a Doppler shift $\nu_j = f_d \cos(\phi_j - \phi_\nu)$, where f_d is the maximum possible Doppler shift, which is given by $f_d = \nu/\lambda$, where ν is the velocity of the mobile and λ is the carrier wavelength. The quantity f_d is also known as the Doppler spread and sometimes also called the fading bandwidth of the channel. By assuming an omni-directional antenna, it has been shown in [43] that the Doppler power spectral density, $S_D(f, \phi_r)$, is given by

$$S_D(f, \phi_r) = \frac{1}{f_d \sqrt{1-(f/f_d)^2}} \cdot \left(S_M(\tau, \phi_\nu + |\cos^{-1}(f/f_d)|) + S_M(\tau, \phi_\nu - |\cos^{-1}(f/f_d)|) \right)$$

$$\forall \quad |f| < f_d \qquad (3.7)$$

where $S_M(\tau, \phi_r)$ is the pdf of the DOA as a function of τ (i.e. the MIP as defined in (3.3)) of the multi-path components at the mobile. If the DOA of the signal at the mobile is uniformly distributed over the interval $(0, 2\pi]$, the Doppler spectrum reduces to the well known expression [43]

$$S_D(f) = \frac{1}{\pi f_d \sqrt{1-(f/f_d)^2}} \qquad \forall \quad |f| < f_d. \qquad (3.8)$$

The equality of (3.6) and (3.8) holds for frequency shifts of f that are in the range $\pm f_d$ about the carrier frequency, f_c, and would be zero outside the range. Equation (3.8) have been shown to match experimental data gathered for a mobile radio channel where the DOA at the mobile is uniformly distributed [115]. For any realistic cellular channel, where the DOA is not uniform, (3.6) should be used. Also, the Doppler spread, f_d, and the coherence time, T_0, are reciprocally related as $T_0 \approx 1/f_d$. The Doppler spread is therefore regarded as the fading rate of the channel.

Let us again consider the GBSBEM scattering environment, with DOA pdf at the base-station given by (3.3). From Figure 3.4 we see that the largest magnitude of $S_D(f, \phi_r)$ vary according to the motion of the mobile. When the mobile is traveling directly to the base station, the Doppler spectrum is a maximum at large frequency offsets, while the maximum move to lower frequency offsets as the motion of the mobile becomes perpendicular to the LOS direction.

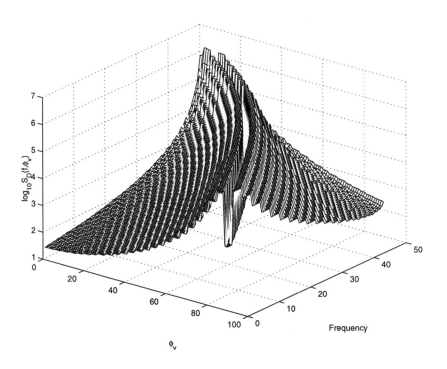

Figure 3.4. $S_D(f, \phi_r)$ as a function of DOA and TOA at base station for the GBSBEM model.

We have seen before that, due to signal dispersion, the coherence bandwidth, $\mathcal{BW}_{ch}$, sets an upper limit on the signalling rate which can be used without suffering frequency-selective fading. Similarly, we have seen that, due to Doppler spreading, the channel fading rate, f_d, sets a lower limit on the signalling rate that can be used without suffering fast fading.

3.2.4 Space-time Correlation Function

The function, $R_D(\Delta t, \phi_r)$ ($\Delta t = t_2 - t_1$) is designated the space-time correlation function at the base station. This quantity is the autocorrelation function of the channel response to a sinusoid and specifies the extent to which there is correlation between the channel's response to a sinusoid sent at time t_1 and the response to a similar sinusoid sent at time t_2. The coherence time, T_0, is

ASPECTS INFLUENCING SPACE-TIME PERFORMANCE

a measure of the expected time duration over which the channel's response is essentially invariant. It is instructive to note that for an ideal time-invariant channel, the channel's response would be highly correlated for all values of Δt, and $R_D(\Delta t, \phi_r)$ would be a constant function.

The relationship between the Doppler power spectrum, $S_D(f, \phi_r)$, and the space-time correlation function, $R_D(\Delta t, \phi_r)$, is given by

$$R(\Delta t, \phi_r) = \int S_D(f, \phi_r) \exp(-jf\Delta t) df, \qquad (3.9)$$

and an example for the GBSBEM scattering environment is shown in Figure 3.5. From the figure is it clear that when the mobile is traveling towards the base station, the coherence time of the channel increases as opposed to the mobile traveling in a direction perpendicular to the LOS direction. Also, for small values of Δt, the correlation between signals received at the base station is high.

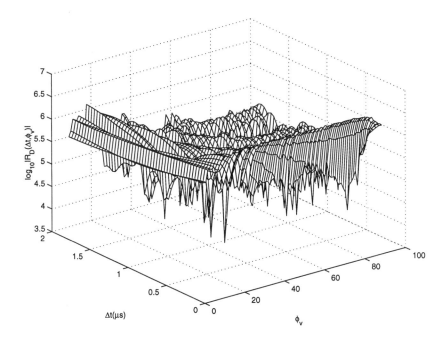

Figure 3.5. $R(\Delta t, \phi_r)$ as a function of DOA and TOA at base station for the GBSBEM model.

Due to the time-variant nature (or fading rate) of the mobile channel, the time variation can be viewed in terms of fast fading and slow fading.

3.2.5 Frequency-selective Fading

The relationship between maximum excess delay time, τ_m, and symbol time, T_s, can be viewed in terms of two different degradation categories, frequency-selective fading and frequency non-selective or flat fading. A channel is said to exhibit frequency-selective fading if $\tau_m > T_s$. This condition occurs whenever the received multi-path components of a symbol extend beyond the symbol's time duration. Such multi-path dispersion of the signal yields inter-path-interference (IPI). In this case, mitigating the distortion is possible because many of the multi-path components are resolvable by the receiver [251, 10].

Frequency-selective fading also occurs if $BW_{ch} < 1/T_s \approx BW$, where $1/T_s$ is normally taken to be equal to the signal bandwidth BW. The signal bandwidth will differ from $1/T_s$ due to system filtering or data modulation type. Distortion occurs whenever a signal's spectral components are not all affected equally by the channel. Some of the signal's spectral components, falling outside the coherence bandwidth, will be affected differently from those components contained within the coherence bandwidth.

3.2.6 Flat Fading

Flat fading occurs under the condition $\tau_m < T_s$. In this case all the received multi-path components of a symbol arrive within the symbol time duration and hence the components are resolvable. Here, there is no channel-induced ISI distortion since the signal time spreading does not result in significant overlap among adjacent received symbols. Performance degradation is still present since unresolvable phasor components can add up destructively to yield a substantial reduction in SNR. For loss in SNR due to flat fading, the mitigation technique called for is to improve the received SNR, or reduce the required SNR. Error control coding is a most effective way to accomplish this.

Another way of looking at flat fading is whenever $BW_{ch} > BW \approx 1/T_s$. Hence, all of the signal's spectral components will be affected by the channel in a similar manner, i.e. fading or no fading. As a mobile changes position, there will be times when the received signal experiences frequency selective distortion even though $BW_{ch} > BW$. In this situation, the null of the channel's frequency transfer function occurs at the center of the signal band. Whenever, this occurs, the DC component of the baseband pulse will be corrupted. One consequence of this is the absence of a reliable pulse peak on which to establish the timing synchronization, or from which to sample the carrier phase carried by the pulse [10]. Thus, even though a channel is categorized as flat fading, based on rms relationships, it can still manifest frequency selective fading on occasions. It is, therefore, apparent that a channel characterized as having flat fading degradation, does not exhibit flat fading all the time. As BW_{ch} becomes much larger than BW (or τ_m much smaller than T_s), less time will be spend in conditions like this. In a CDMA system, since T_c is very small this condition can become a practical problem. By comparison, it should be clear that for frequency-selective fading, the fading is independent of the position of the signal band, and frequency-selective fading occurs all the time.

3.2.7 Fast Fading

Fast fading is used to describe channels in which $T_0 < T_s$ and describes a condition where the time duration in which the channel behaves in a correlated manner is short compared to the time duration of a symbol. Therefore, it can be expected that the fading character of the channel will change several times while a symbol is propagating, leading to distortion of the baseband pulse shape. As a result, fast fading distorts the baseband pulse, resulting in a loss of SNR that often yields an irreducible error rate.

Fast fading occurs if the symbol rate is less than the fading rate, stated as $\mathcal{BW} < f_d$. This definition is strictly correct, but to prevent random frequency modulation (FM) due to varying Doppler shifts, the condition $\mathcal{BW} \gg f_d$ should be adhered to, which is normally the case in WCDMA systems such as UMTS.

3.2.8 Slow Fading

Slow fading refers to channels in which $T_0 > T_s$. For this condition, the time duration in which the channel behaves in a correlated manner is long compared to the time duration of the transmission symbol. As a result, it can be expected that the channel state will remain constant during the time in which a symbol is transmitted. As with flat fading, the main degradation in a slow fading channel is a loss in SNR.

Slow fading occurs if the signalling rate is greater than the fading rate, stated as $\mathcal{BW} > f_d$.

3.3 SCATTERING ENVIRONMENT

As has been mentioned before, the distribution of the DOA of multi-path signals is often assumed to be uniform over $(0, 2\pi]$ [192]. To determine the performance of a space-time system, channel models that include the effect of the DOA need to be constructed. A critical aspect that determines the DOA at either the base station or the mobile, is the scattering environment around the transmitter and receiver. More detail will be given about the various channel models and scattering environments in Chapter 4. In what follows, space-time channels are defined as being either high-rank or low-rank to include angular dispersion and build on the previous definition of delay spread. The classification of space-time channels presented here is taken from [83].

A very important point to note is that, due to the difference in angular dispersion at the base station and mobile station, a channel may be high rank in the downlink, but low rank in the uplink. This is an important difference when comparing space-time channels to purely temporal channels.

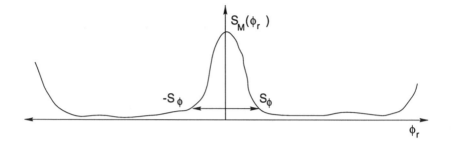

Figure 3.6. Unfolded pdf of received signal power as a function of ϕ_r.

3.3.1 High-rank Channels

A channel is said to be high rank if τ_m is equal to or larger than T_s, or the angular spread, S_ϕ, is equal to or larger than the 3 dB beam-width ϕ_{3dB} of the antenna pattern. In other words, a channel is high rank when

$$\tau_m \geq T_s \quad \text{or} \quad S_\phi \geq \phi_{3dB}, \tag{3.10}$$

where the angular spread, S_ϕ, is defined as the square root of the second central moment of the unfolded angular power distribution [60] or

$$S_\phi = \int_{\phi_r-180}^{\phi_r+180} (\phi_r - \phi_0)^2 \, S_M(\phi_r) d\phi_r. \tag{3.11}$$

A physical interpretation of S_ϕ is shown in Figure 3.6, and from the figure it is clear that the scattering environment will influence the way in which the base station receives signals from the mobile. Specifically, from this definition, it is clear that a high-rank channel has its characteristics due to large delay spreads and/or large angular spreads. High-rank channels should, therefore, take into account local scattering together with distant scatterers. This concept is shown in Figure 3.7, where typical pdf's of the DOA of signals at the base station is shown for various cellular environments. The common assumption of a uniform pdf of the DOA is closely approximated by a pico cell with no LOS component present. Clearly this type of channel is not the predominant channel used in cellular systems, and the great impact of the channel model chosen on the DOA is clearly visible.

3.3.2 Low-rank Channels

A channel is said to be low rank if τ_m is smaller than T_s, and the angular spread S_ϕ is small compared to the 3 dB beam-width of the antenna pattern. In other words, a channel is low rank when

ASPECTS INFLUENCING SPACE-TIME PERFORMANCE 57

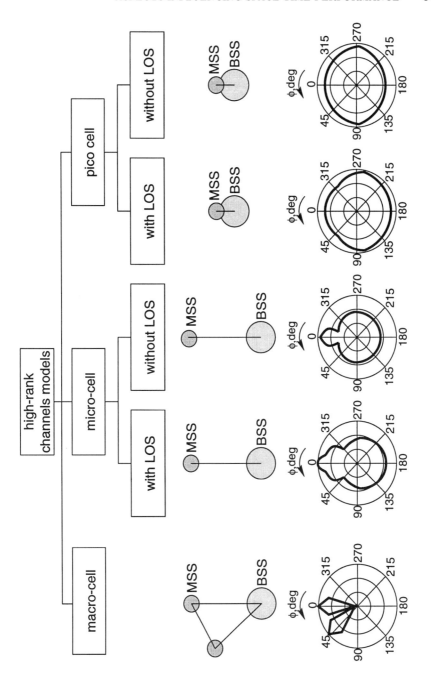

Figure 3.7. Channel models for high-rank channels.

$$\tau_m \ll T_s \quad \text{and} \quad S_\phi \ll \phi_{3dB}. \tag{3.12}$$

A low-rank channel requires that contributions from scatterers around the base station and from distant scatterers are negligible and consists of the local scattering model around the mobile associated with a LOS component. This scenario is shown in Figure 3.8 and from the figure it is clear that the modeling of the scattering environment is less important than in the case of a high-rank channel. The range for the radius of the scatterer is in the order of 50 to 300m, with maximum radius that of a macro-cell. The azimuthal power spectrum is the mean value of the incident power of the various multi-path components. From the definition of a low-rank channel, it is clear that this angular distribution cannot be resolved by the antenna main lobe. However, DOA estimation algorithms can still determine angles of arrival with an accuracy that is much higher than the beam-width of the antenna. The actual resolution accuracy is mainly limited by the angular spread around the mean direction of arrival. In space-time systems we are sometimes interested in putting antenna pattern nulls on interferers. The nulls are much sharper than the main lobes, and the amount of residual interference is critically determined by the product of the antenna pattern near the null and the azimuthal power spectrum around the direction of the interferer.

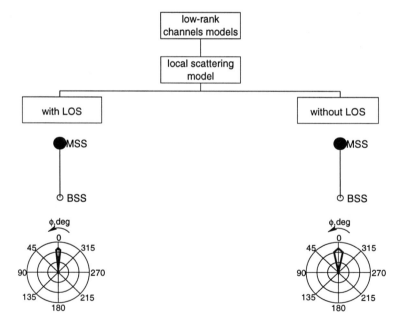

Figure 3.8. Channel models for low-rank channels.

ASPECTS INFLUENCING SPACE-TIME PERFORMANCE   59

3.4 ANGULAR SUBSCRIBER DISTRIBUTION

The distribution of the DOA of signals in a cellular system is dependent on both the distribution of subscribers in a cell, and the distribution of scattering elements around each subscriber. Specifically, the manner in which subscribers are clustered together in angle (as would be the case on a road), significantly influences the gains that may be achieved by a space-time system. For instance, if the reference user and an interfering user are co-located in angle, no antenna pattern can be formed in either the up- or downlinks to reduce the interference experienced by the reference user. Therefore, the gain offered by, for instance, a SFIR system to users in the relevant cell is negligible. On the other hand, having subscribers clustered in certain areas means that antenna sectors can be narrowed, thereby reducing interference to adjacent cells and increasing the overall network performance, even if the performance of all individual cells is not increased.

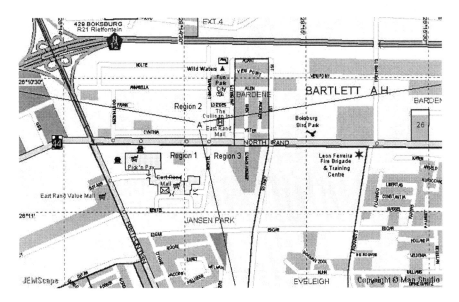

Figure 3.9.   Typical practical implementation scenario of a cellular network.

When planning a cellular wireless access network, be it a mobile or wireless local loop network, one of the first steps that the network planner takes is the estimation of the traffic demand in various areas. Consider the case shown in Figure 3.9. This figure depicts a map of a busy shopping mall area in Boksburg, South Africa. Assume for instance that due to land acquisition considerations, a base station is placed at location A. The traffic that can be expected by this base station is clearly a function of angle. In region 1, high levels of traffic are generated by users of mobile phones in the mall. These users have low mobility and the frequency of calls may be high. In region 2, however, the base station mostly sees traffic from mobile users traveling at speeds of up to 120 km/h with short call durations, a typical highway scenario. Finally in region 3, the base station will experience low levels of traffic as residential dwellings are mainly covered by the base station in this area. Based on this real-world scenario, it is clear that assuming a uniform distribution of users in a cell may, in a number of circumstances, not be valid. The above-mentioned rationale is further confirmed by the results in [278], where it is shown that geographic and demographic factors significantly influence all aspects of a cellular network.

Therefore, a number of approaches to the modeling of user locations have been followed in literature. For example, in [110] a uniformly distributed mobile user density is proposed, with a highway traffic distribution model presented in [145] and a modified Gaussian distribution proposed in [204]. The most accurate description of user locations would be gained from practical measurements at each site of interest. These assumptions are sufficient in the intended environments. A more general pdf that describes the user distribution, and is applicable to many scenarios, has been proposed in [157] and is discussed below.

For this pdf, it is assumed that the cellular multiple access system is constituted by a number of non-overlapping hexagonal cells, each of radius R_r. Each cell has a base station located at its center, with mobile users distributed throughout the cell. The position of a mobile user in the cellular structure (reference and adjacent cells) is fully defined by its distance R from the reference base station, and its angle, ϕ_0, measured as shown in Figure 3.1. In general then, the pdf of the angular distribution of users in a cellular network can be expressed as

$$p_{\Phi_0}(\phi_0) = \frac{1}{A_{norm}} \left[1 + \sum_{l=1}^{N_{peak}} \varrho_l \left[\text{rect}\left(\frac{\varpi_l \phi_0}{\pi} - \varsigma_l\right) \right. \right.$$
$$\left. \left. + \text{rect}\left(\frac{\varpi_l \phi_0}{\pi} - \varsigma_l - 2\pi\right) \right] \cos^2(\varpi_l \phi_0 - \varsigma_l) \right] \quad 0 \leq \phi_0 \leq 2\pi, \quad (3.13)$$

where A_{norm} is a normalizing factor to ensure that $\int_0^{2\pi} p_{\Phi_0}(\phi_0)d\phi_0 = 1$, and N_{peak} is the number of peaks in the pdf. The factor N_{peak} is a measure of the angular clustering of mobiles in a cell. Clearly, if $N_{peak} = 0$, (3.13) denotes a uniform angular subscriber distribution. Furthermore,

$$\text{rect}(x) = \begin{cases} 1 & |x| < \frac{1}{2} \\ 0 & |x| > \frac{1}{2} \end{cases}, \quad (3.14)$$

with ϖ_l an integer controlling the width of peak l. Typically, values for ϖ_l can be chosen to yield angular peaks of different maximum and minimum widths. For instance, if $\varpi_l = 10$ the width of peak l in the pdf of the angular distribution of mobiles will be 18°. The angular location of peak l is given by ς_l, with the relative size of each peak determined by ϱ_l (typically values between 0 and 100).

As an example, consider the scenario depicted in Figure 3.10. This figure depicts a hypothetical urban or suburban scenario where two main roads intersect with a base station situated at the crossing. As these roads carry a much larger portion of the total mobile traffic than other areas within the coverage region of the cell, the probability of receiving transmission with angles of incidence of approximately 0°, 90°, 180° and 270° is substantially higher, as the concentration of mobiles in these specific directions is high. The pdf of the location of mobiles for this scenario is shown in Figure 3.11, where the angular width of a street is taken to be 10°.

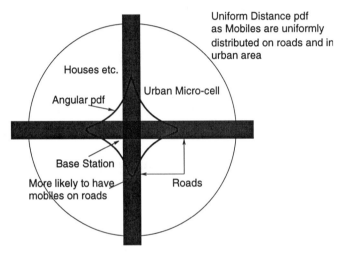

Figure 3.10. Qualitative description of the pdf of the mobile distributions in a Typical Urban micro-cellular scenario.

Furthermore, by using the variable transformation

$$R = \phi_0 \frac{3\,R_r}{2\pi}, \qquad (3.15)$$

where R_r is the cell radius, (3.13) can also describe the distance distribution of mobiles. However, a uniform distance distribution is often a good approximation to make (see for instance Figure 3.10). Using the above equations, different scenarios, such as the clustering of mobiles, can also be represented.

3.5 SUMMARY

Path loss, fading, scattering environment and user distribution are some of the key aspects limiting the performance of space-time processing systems and is crucial in determining mitigation techniques. This chapter described and

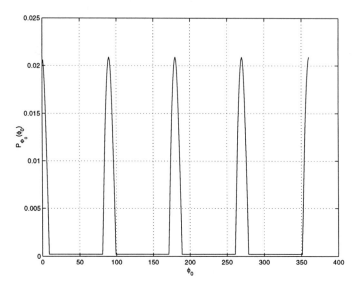

Figure 3.11. Example pdf of the angular distribution of mobiles.

summarized these performance limiting aspects. Chapter 5 will consider smart antenna techniques for mitigating these effects. The next chapter will consider channel models to describe the space-time environment more effectively.

4 SPACE-TIME CHANNEL MODELS

When realistic channel models of a mobile communication system are available, efficient signal processing schemes can be devised to improve system performance, and accurate system analysis can also be performed to predict system capacity and performance. In general, models describe parameters such as received signal strength, power delay profiles and Doppler spectra, which are important for the analysis of systems with omni-directional antannas. Of high importance in space-time systems, is knowledge of the direction of arrival (DOA) of the received signals, which is not available from conventional models. Using the classical understanding of concepts such as fading, Doppler spread, correlation, etc., spatial channel models can be built to incorporate concepts such as time delay spread, DOA and adaptive array geometries to mention only a few. In this chapter, channel models are reviewed and discussed. Very important effects are described such as multi-path fading, and models for the scattering surrounding the mobile and base station. It is of importance to note that, due to the difference in angular dispersion at the mobile and base station, the propagation characteristics in the uplink and downlink might be different, and this is of significance in space-time based system performance analysis.

4.1 BASIC MODELING

In a wireless mobile communication system, a signal can travel from transmitter to receiver over multiple reflective paths and this is called multi-path propagation. This effect can cause fluctuations in the received signal's amplitude, phase and angle of arrival, giving rise to what is called multi-path fading.

The basic structure of a space-time based multiple access communication system is shown in Figures 2.9 and 2.10. It is assumed that the system will consist of K users, each transmitting a signal over an independent discrete multi-path channel with L propagation paths to the receiver, each with its own amplitude, phase and DOA. The distribution of these parameters is dependent on the type of environment (i.e. macro-, micro- or pico cell).

The DOA component is a function of three different aspects namely

1. scattering at the mobile (this scattering is normally also influenced by the mobile speed);

2. scattering at the base station; and

3. distant scatterers. This type of scattering can occur in urban and rural environments due to large structures, such as mountains, buildings, etc., and has an influence on the mobile channel even though the scatterers are far away from the base station and mobile. If these structures have LOS to both the base station and mobile, they can act as discrete reflectors or as clustered reflectors. When the reflectors are clustered, the base station or mobile antenna see the scattered components as in points 1 and 2 above.

It is clear that modeling of the scattering environment plays a crucial role in system design. Various models for the scattering environment have been proposed, with different properties and with varying degrees of accuracy. In the next sections, these scattering models are described.

4.2 LOCAL SCATTERING ENVIRONMENT

In classical channel models for mobile communications, the distribution of the DOA of multi-path signals has been assumed to be uniform over $(0, 2\pi]$ [43]. Since system performance in a smart antenna system is directly proportional to the DOA of the multi-path components, more realistic DOA models are needed. This section summarizes spatial scattering environments proposed in the literature for use in spatial channel models. Emphasis is placed on scattering models relevant to a CDMA environment, and therefore, GSM-based simulation models such as the TU model [332, 333, 94] and BU model [332, 333, 94] will not be described here. For a detailed discussion of spatial scatterer models, the text by Ertel et al. [64] can also be consulted.

For the circular disk of scatterers model (CDSM) proposed originally by Jakes [115] and used in Lee [140], and the Gaussian scatterer (GS) model [159], a complete derivation of the DOA and correlation is given. These two models have been chosen since the CDSM is the "classical" spatial model, while the GS model is more realistic in certain environments. The derivation of these two models is also included to give the reader the necessary background to calculate DOA models for other scattering environments as the need arises.

Local scattering models have been developed and are used for different applications. Some were intended to provide information about a single channel characteristic, while others attempt to capture all the properties of the wireless channel. Most of the models are based on measurements for a specific environment. For instance, measurements reported in [83] suggest that the DOA statistic in rural and suburban environments is Gaussian distributed and forms the basis for the Gaussian angle of arrival (GAA) model. It has further been shown in [312] that scattering elements can best be described by a Gaussian bell shape around the mobile in an urban environment. In this case it is assumed that the majority of scattering points are clustered together, with the density of scattering points decreasing as the distance from the mobile is increased. It is especially in urban micro-cellular environments that this assumption provides a good approximation of the physical reality. It will be seen that the GS model gives a DOA distribution that is close to the GAA model.

Although these models are based on measurements, the validity of the mathematical model is always in question. Using the CDSM, it was reported in [64], that this model had been used to simulate a Rayleigh fading spatial channel. The error rate when using a $\pi/4-$ DQPSK system, was compared with measurements taken in a typical suburban environment, with predicted error rate estimates within a factor of two of the actual measured error rate, indicating a reasonable degree of accuracy for the model. As is discussed in [64], theoretical modeling of the DOA should be done carefully in order to approximate reality. For instance, in practice, the DOA will be discrete and therefore it is not always valid to use a continuous DOA distribution to estimate the correlation present between the antenna array elements. Correlation effects from a continuous DOA pdf decrease monotonically as a function of antenna spacing, whereas correlation properties resulting from a discrete DOA pdf have damped oscillations present. The result of this is that a continuous DOA distribution will underestimate the correlation between the antenna elements in the array.

4.2.1 Circular Disk of Scatterers Model (CDSM) [115, 140, 284, 232]

Making extensive use of the results by Van Rheeder *et al.* [284], the DOA pdf at the base station is derived here for the CDSM shown in Figure 4.1.

If the mobile is located at the co-ordinates $(0,0)$, then the pdf of the location of scatterers around the mobile is given by

$$p(x,y) = \begin{cases} \frac{1}{\pi R_D^2}, & x^2 + y^2 \leq R_D^2 \\ 0, & \text{elsewhere} \end{cases}. \tag{4.1}$$

Finding the joint density $p_{R_s,\Phi}(r_s,\phi)$ requires a transformation of the random variables (X, Y) into the random variables (R_s, Φ) via

$$p_{R_s,\Phi}(r_s,\phi) = \mid J \mid p_{X,Y}(x,y) \mid_{x=r_s \cos\phi - R, y=r_s \sin\phi}. \tag{4.2}$$

In (4.2), J is the Jacobian of the transformation, computed as

66 SPACE-TIME PROCESSING FOR CDMA

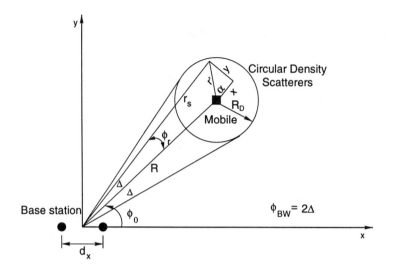

Figure 4.1. Circular disk of scatterers model geometry.

$$J = \begin{vmatrix} \frac{\partial(r_s \cos\phi - R)}{\partial r_s} & \frac{\partial(r_s \sin\phi)}{\partial r_s} \\ \frac{\partial(r_s \cos\phi - R)}{\partial \phi} & \frac{\partial(r_s \sin\phi)}{\partial \phi} \end{vmatrix} = r_s. \quad (4.3)$$

Substituting (4.1) and (4.3) into (4.2) yields the joint density of R_s and Φ

$$p_{R_s,\Phi}(r_s,\phi) = \begin{cases} \frac{r_s}{\pi R_D^2}, & r_s^{(1)} \leq r_s \leq r_s^{(2)}, \phi^{(1)} \leq \phi \leq \phi^{(2)} \\ 0 & \text{otherwise} \end{cases} \quad (4.4)$$

where

$$\begin{aligned}
r_s^{(1)} &= \sqrt{R_D^2 + R^2(1 - 2\sin^2\phi) - \mathcal{B}} \quad (4.5) \\
r_s^{(2)} &= \sqrt{R_D^2 + R^2(1 - 2\sin^2\phi) + \mathcal{B}} \\
\mathcal{B} &= 2R\cos\phi\sqrt{R_D^2 - R^2\sin^2\phi} \\
\phi^{(1)} &= -\sin^{-1}(1/\upsilon) \\
\phi^{(2)} &= \sin^{-1}(1/\upsilon) \\
\upsilon &= \frac{R}{R_D}.
\end{aligned}$$

Limits on the parameter r_s were determined by fixing ϕ and then computing the points at which the resulting line intersected the scattering circle in Figure

4.1. Limits on ϕ were determined by finding the angles of the two tangent lines connecting the scattering circle with the base station.

Integrating (4.4) with respect to r_s, gives the desired DOA density

$$p_{\Phi,\nu}(\phi,\nu) = \begin{cases} \frac{1}{2\pi} \left(\nu \cos\phi + \sqrt{1 - \nu^2 \sin^2\phi} \right)^2; & 0 \leq \phi \leq 2\pi,\ 0 \leq \nu \leq 1 \\ \frac{2}{\pi} \nu \cos\phi \sqrt{1 - \nu^2 \sin^2\phi}; & \phi^{(1)} \leq \phi \leq \phi^{(2)};\ \nu > 1 \\ 0 & \text{otherwise.} \end{cases}$$
(4.6)

This result is valid for $\nu > 1$, i.e., when the base station is located outside the scattering circle and when the base station is located inside the scattering circle, $0 \leq \nu \leq 1$.

Figure 4.2 depicts $p_{\Phi,\nu}(\phi,\nu)$ as a function of both ϕ and ν. From this figure, the following observations can be made. For large ν, the situation where the mobile terminal is far from the base station, and all the scatterers are close to the mobile, the pdf approaches an "impulse-like" density. Conversely, for small ν, the situation where the mobile is close to the base station and the scattering circle is large (NLOS propagation), the pdf approaches a uniformly distributed density. Using (4.6) and setting ν equal to zero, the DOA density is calculated as $1/2\pi$.

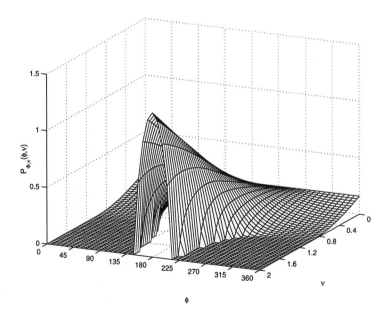

Figure 4.2. DOA as a function of ϕ and ν.

4.2.1.1 Correlation.
In the CDSM it is assumed that all signals arrive at the base station within $\pm\Delta$ of the angle ϕ_0, and that the ith received signal path is uniformly distributed with height $1/2\Delta$. Using the results of Lee [139], the fading correlation between two antenna elements, spaced d_x apart, can be written as two components R_{xx} and R_{xy}

$$R_{xx} = \int_{-\pi/2+\phi_0}^{+\pi/2+\phi_0} \cos[\,2\pi(d_x/\lambda)\sin\phi\,]\; p_\Phi(\phi)\, d\phi \qquad (4.7)$$

$$R_{xy} = \int_{-\pi/2+\phi}^{+\pi/2+\phi_0} \sin[\,2\pi(d_x/\lambda)\sin\phi\,]\; p_\Phi(\phi)\, d\phi, \qquad (4.8)$$

where R_{xx} denotes the correlation of the real components of the signal received at the two antennas, and R_{xy} denotes the correlation of the real component of the signal arriving at the one antenna and the imaginary component arriving at the other antenna.

From (4.6) with v constant, the density $p_\Phi(\phi)$ is used to derive R_{xx} and R_{xy} as series expansions of integer order Bessel functions. Substituting (4.6) into (4.7) and (4.8), and using geometric substitutions and some numerical analysis yields the correlation of fading approximations [284]

$$R_{xx} \approx \left[J_0\left(\frac{2\pi d_x}{\lambda v}\cos\phi_0\right) + J_2\left(\frac{2\pi d_x}{\lambda v}\cos\phi_0\right) \right] \cos\left(\frac{2\pi d_x}{\lambda}\sin\phi_0\right), \qquad (4.9)$$

and

$$R_{xy} \approx \left[J_0\left(\frac{2\pi d_x}{\lambda v}\cos\phi_0\right) + J_2\left(\frac{2\pi d_x}{\lambda v}\cos\phi_0\right) \right] \sin\left(\frac{2\pi d_x}{\lambda}\sin\phi_0\right), \qquad (4.10)$$

where $v \gg 1$. These approximations lead to a simple expression for the envelope correlation

$$\begin{aligned}\rho &= \sqrt{R_{xx}^2 + R_{xy}^2} \qquad (4.11)\\ &\approx \left| J_0\left(\frac{2\pi d_x}{\lambda v}\cos\phi_0\right) + J_2\left(\frac{2\pi d_x}{\lambda v}\cos\phi_0\right) \right|.\end{aligned}$$

Figure 4.3 depicts the fading correlation envelopes for the CDSM DOA when $\phi_0 = 45°$. In Figure 4.3, increasing values of Δ denote larger scattering areas as would be found in micro-cells with NLOS propagation. Thus, whereas $\Delta = 40°$ may be used to represent a NLOS micro-cell, a value of $\Delta = 10°$ may be used to represent a macro-cell. From the figure it is then clear that the CDSM indicates that larger antenna spacing is required in macro-cellular environments to decorrelate signals received by a diversity receiver. Conversely, this would mean that with fixed antenna spacing, larger diversity gains could be achieved in an environment where severe scattering is present than in an environment where few scattering points are present.

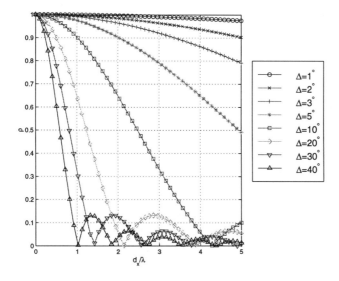

Figure 4.3. CDSM fading correlation for $\phi_0 = 45°$.

4.2.2 Gaussian Scatterer (GS) Model [312, 83, 159, 158]

In the previous section the classic CDSM was used to describe the local scattering elements surrounding a mobile. This model provides a concise mathematical framework to use for the description of local scattering elements. However, in [312, 83] it is shown that scattering elements can be best described by a Gaussian bell shape as shown in Figure 4.4. In this case it is assumed that the majority of scattering points are clustered together, with the density of scattering points decreasing as the distance from the mobile is increased. It is especially in micro-cellular environments that this assumption is a good approximation of physical reality. For instance, standing in an urban area a mobile would see a large number of buildings (i.e. possible sources of scattering) in its immediate vicinity. Also, a "second tier" of buildings would be visible when looking between the surrounding buildings and so forth. As the distance from the mobile is increased, fewer and fewer buildings will be visible as they are obscured by closer structures. Furthermore, consider the MIP, $S_M(\tau, \phi_r)$, recommended in COST-207 for Rural, Typical Urban and Bad Urban scenarios [45] and Appendix C. In all three cases, the intensity profiles have exponential shapes. Coupling this recommendation with the assumption that all delayed paths are caused by local scattering elements, the Gaussian distribution of local scattering elements make intuitive sense as the mobile station would see a lot of possible scattering points in its immediate vicinity (small delay values), and fewer scattering points would be visible at larger distances (bigger delay values). Physical measurements in urban micro-cellular areas, such as those performed in the TSUNAMI project [60], confirm this train of thought.

The argument is, however, not true in the case of hilly terrain and indoor environments. In these cases, the mobile is surrounded by very pronounced geographical elements (the hills or the walls of the office) and the assumption of a ring of scattering elements may be more appropriate. As an example, measured power delay profiles in an indoor environment [60] can be compared to those in micro-cellular environments. Therefore, by using the GS model the angular distribution distribution of the signals arriving at the base station can be calculated as an alternative to the CDSM.

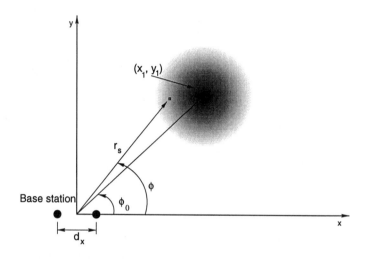

Figure 4.4. Modeling of scattering elements using the GS model.

Consider the general case of a scattering area situated at a point (x_1, y_1) in space, with the base station located at the origin as shown in Figure 4.4. In this case, the density of the scattering elements at a distance r_s and angle ϕ from the mobile, can be described by the bivariate Gaussian distribution [312]

$$\begin{aligned} p_{R_s,\Phi}(r_s,\phi) &= \frac{1}{2\pi\sigma_s^2} \exp\left(-\frac{(x-x_1)^2 + (y-y_1)^2}{2\sigma_s^2}\right) \\ &= \frac{1}{2\pi\sigma_s^2} \exp\left(-\frac{r_b^2 - 2r(x_1\cos\phi + y_1\sin\phi) + x_1^2 + y_1^2}{2\sigma_s^2}\right), \end{aligned} \quad (4.12)$$

where ϕ and r_s are shown in Figure 4.4 and, σ_s the standard deviation of the local scattering elements.

In order to calculate the pdf of the DOA of signals at the base station caused by reflections from the scattering area, the cumulative distribution of scattering elements as a function of ϕ is required. This cumulative distribution, or the addition of more scattering elements as a function of angle can be expressed as

$$W_\phi = \int_{-\pi}^{-\pi} \int_0^\infty p_{R_s,\Phi}(r_s,\phi) dr_s d\phi. \tag{4.13}$$

Then, the probability of receiving a signal with a certain DOA is directly related to the density of scatterers in the specific direction, or the derivative of (4.13) with respect to ϕ. Therefore

$$p_\Phi(\phi) = \frac{d}{d\phi} W_\phi = \int_0^\infty p_{R_s,\Phi}(r_s,\phi) dr_s. \tag{4.14}$$

Using standard tables of integrals [2] and algebraic manipulation, (4.14) can be simplified to

$$\begin{aligned}p_\Phi(\phi) &= \frac{1}{2\sqrt{2\pi}\sigma_s} \exp\left(\frac{x_1^2(\cos^2\phi - 1) + y_1^2(\sin^2\phi - 1) + 2x_1 y_1 \cos\phi \sin\phi}{2\sigma_s^2}\right) \\ &\quad \cdot \operatorname{erfc}\left(\frac{-(x_1\cos\phi + y_1\sin\phi)}{\sqrt{2}\sigma_s}\right),\end{aligned} \tag{4.15}$$

where

$$\operatorname{erfc}(x) = \frac{2}{\sqrt{\pi}} \int_z^\infty e^{-t^2} dt, \tag{4.16}$$

is the well known complementary error function [2].

In order to use (4.14) in calculations, the standard deviation of the local scattering elements needs to be known. It should be clear that the standard deviation of the local scattering elements has to be determined for each specific case (physical cell geometry) if accurate values are required. This is very difficult and can only be done using ray-tracing methods or extensive high resolution measurements [176, 30]. However, in [159], a general rule of thumb for the first order approximation of the standard deviation is formulated.

As a basis for the derivation of the rule, consider again the MIP recommended in COST-207. It can be assumed that the stronger multi-path echoes that arrive at the base station shortly after the LOS signal are due to scattering points close to the mobile (shorter transmission path length and hence lower path loss factors), and that multi-path echoes with larger delay values are due to scattering points further from the mobile. In the case of a TU area, the intensity of multi-path signals will have dropped by half after 0.69 μs [45]. This delay value translates to a path length difference of 207 m. Typical micro-cells with LOS propagation have cell radii in the order of 1 km [97], yielding a typical ratio of path length difference to cell radius of approximately $0.2R_r$, where R_r is the radius of the cell in meters. This value may be used as the standard deviation for the Gaussian distribution of the local scattering elements under LOS micro-cellular conditions.

Shifting the focus to NLOS micro-cellular environments, the mobile signal may rely only on reflected signals to reach the base station as a large number of reflection points are found in these environments. Consequently the standard deviation of the scattering elements should be larger. On the other hand, in macro-cellular scenarios each mobile user sees a substantially smaller proportion of the scattering elements in the cell as the base station and the mobile are separated by a much larger distance. Based on empirical evidence and standard channel models, a rule of thumb for the standard deviation of the Gaussian bell shape described in (4.12) can then be formulated as $\sigma_s = 0.34 R_r$ for NLOS micro-cellular conditions; $\sigma_s = 0.2 R_r$ for LOS micro-cellular conditions and $\sigma_s = 0.1 R_r$ for macro-cellular conditions (see Table 4.1).

	Cell Radius	Standard Deviation
LOS Micro-cell	0.4 - 2 km	$0.2\, R_r$
NLOS Micro-cell	0.4 - 2 km	$0.34\, R_r$
Macro-cell	2-20 km	$0.1\, R_r$

Table 4.1. Cell radii and local scattering element standard deviations for macro- and micro-cellular environments.

The accuracy of the rule of thumb can be tested against measurement results. Clearly, if site specific propagation data is available, it is possible to calculate more exact values for the local scatterer variance. In [312] it is shown that the cross-correlation of the fading envelopes of multi-path signals, generated due to a Gaussian distributed scattering area, received by antenna elements spaced by d_x meters is

$$\rho(d_x) = e^{-\frac{4\pi d_x^2 \sigma_{angular}^2}{\lambda^2}}, \qquad (4.17)$$

where λ denotes the wavelength of the carrier frequency and $\sigma_{angular}^2$ denotes the variance of the angular spread of signals arriving at the base station antenna. From (4.17) and the well known formula for the length of an arc, the normalized (with respect to distance from the base station) standard deviation of the scattering elements can be calculated as

$$\sigma_s = \sqrt{\frac{\lambda^2 \ln(\rho(d_x))}{-4\pi d_x^2}}. \qquad (4.18)$$

Using (4.18) and correlation results presented in [183] (see also Figure 4.5), the local scatterer variance for the case of a rural macro-cell can be calculated as $0.113\, R_r$, and for the case of a NLOS micro-cell as $0.337\, R_r$, both of which correspond well with the rule of thumb presented above.

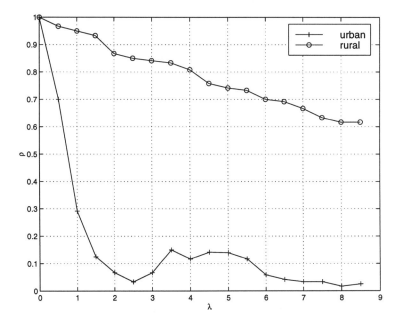

Figure 4.5. Envelope correlation for rural macro- and urban micro-cells [183].

Furthermore, consider Figure 4.6 where the pdf of the DOA of signals at the base station calculated using the rule of thumb is shown. For both microcellular cases, R_r was set equal to 1 000 m, and for the macro-cellular case, $R_r = 10000$ m. In Table 4.2, the total spread of the DOA of signals at the base station is compared to typical values for the angular widths of signals seen at the base station found in [194, 95]. In [95, 194], no distinction is made between LOS and NLOS environments and no indication of the cell radius is given. Furthermore, a typical angular spread of 120° is given for dense urban micro-cells. This value is somewhere in the middle of the results obtained using the rule of thumb for the LOS and NLOS scenarios. Furthermore, the values obtained in the two cases for macro-cells are quite close. Based on these results, it can be concluded that the proposed values for the standard deviation of the scattering environment are good approximations of the true values.

Comparing the pdf of the DOA of signals at the base station predicted by the GS model and the CDSM, it is clear that both models yield similar results (see Figure 4.2 and Figure 4.7). In Figure 4.7, the pdf of the DOA of signals at the base station is shown for various values of v (equivalent to v in Figure 4.2). For large values of v (where the mobile terminal is far from the base station and all the scatterers are close to the mobile), the pdf approximates an "impulse-like" density as v increases. Conversely, for small values of v (where the mobile is close to the base station and the scattering circle is large), the pdf approaches a uniformly distributed density.

74   SPACE-TIME PROCESSING FOR CDMA

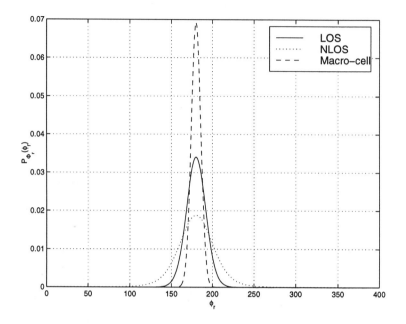

Figure 4.6. Distribution of the DOA of signals at the base station using rule of thumb calculation.

Environment	Rule of thumb	Literature [95, 194]
NLOS Micro-cell ($R_r = 1000$ m)	160°	120°
LOS Micro-cell ($R_r = 1000$ m)	95°	120°
Macro-cell ($R_r = 10000$ m)	38°	20°

Table 4.2. Comparison of angular spread calculated using rule of thumb and typical measured angular widths.

4.2.2.1 Correlation. With reference to Section 4.2.1.1, it is possible to calculate the envelope correlation for the GS model by substituting (4.14) in (4.9) and (4.10) and solving numerically. Results for the envelope correlations as a function of antenna spacing for various scenarios are depicted in Figure 4.8. From the figure it is clear that the GS model and CDSM exhibit similar correlation characteristics with areas having more scattering elements (NLOS micro-cells or large Δ) having a greater decorrelating effect than areas with few scatterers (macro-cells or small Δ). This would indicate that space-time systems implementing diversity would require larger antenna spacings in macro-cells than in micro-cells.

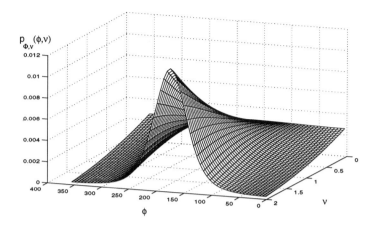

Figure 4.7. DOA as a function of ϕ and v.

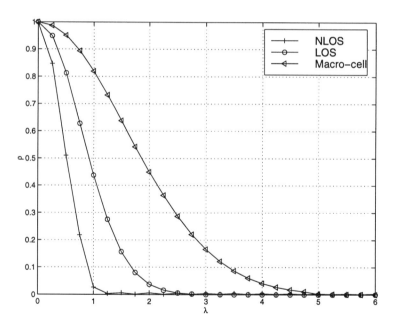

Figure 4.8. GS model fading correlation.

4.2.3 Effective Scatterer Model (ESM) [142, 64]

For this model the scattering elements around the mobile can be described as a ring of scatterers, evenly spaced around the mobile as illustrated in Figure

4.9. It is assumed that N_s scatterers are uniformly distributed on the circle with radius R_D and oriented in such a way that a scatterer is located in the LOS path of the mobile to base station, which is separated by a distance R. The model has application to any arbitrary array size to predict the correlation between the signals received by two elements as a function of element spacing. The discrete DOAs predicted by the model are given by

$$\phi_i \approx \frac{R_D}{R} \sin\left(\frac{2\pi}{N_s} \cdot i\right) \quad \forall\, i = 0, 1, \cdots, N_s - 1. \tag{4.19}$$

From the discrete DOAs, the correlation between any two elements of an array can be found using

$$\rho = \frac{1}{N_s} \sum_{i=0}^{N_s-1} \exp(-j2\pi d_x \cos(\phi_0 + \phi)). \tag{4.20}$$

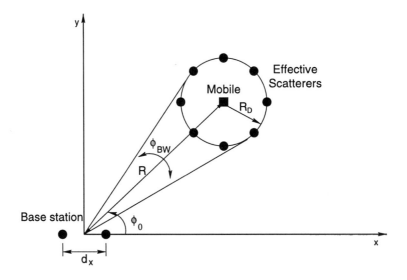

Figure 4.9. ESM geometry.

4.2.4 Discrete Uniform Distribution (DUD) Model [13]

This model is similar to the ESM in terms of motivation and analysis [13]. The model has the following characteristics and properties

- N_s scatterers are evenly spaced over a DOA range within a narrow beam width centered about the LOS to the mobile (Figure 4.10).

- The discrete possible DOA's are given by (N_s odd)

$$\phi_i = \frac{i}{N_s - 1} \cdot 2\Delta \quad \forall \quad i = -\frac{N_s - 1}{2}, \cdots, \frac{N_s - 1}{2}, \quad (4.21)$$

- Predicts correlation coefficient using a discrete DOA model. For two antenna elements with separation d_x, the correlation is given by

$$\rho = \frac{1}{N_s} \sum_{i=-\frac{N_s-1}{2}}^{\frac{N_s-1}{2}} \exp(-j2\pi d_x \cos(\phi_0 + \phi_i)) \quad \forall \quad N_s \quad \text{odd.} \quad (4.22)$$

- Correlation predictions fall off more quickly than the ESM predictions.

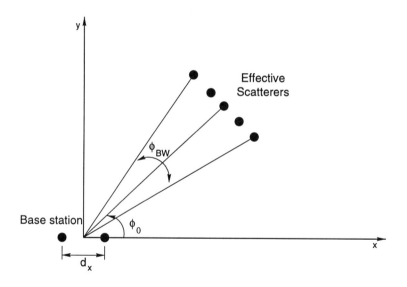

Figure 4.10. DUD model geometry.

4.2.5 Geometrically Based Single-Bounce Statistical Models (GBSB)

GBSB statistical models are defined by spatial scattering density function and are useful for both simulation and analysis purposes. For simulation purposes, the model involves randomly placing scatterers in the scattering region according to the spatial scatterer distribution. From the scatterer location, the DOA, time-of-arrival (TOA) and signal power are determined. It is also possible to derive the joint and marginal DOA and TOA pdf using this model. In this

78   SPACE-TIME PROCESSING FOR CDMA

summary, two models based on the GBSB framework will be discussed. These are the Geometrically Based Circular Macro-cell Model and the Geometrically Based Elliptical Micro-cell Wideband Model. The accuracy of these models is dependent on the shape and size of the spatial scatterer density function, which should ideally be based on measurements.

4.2.5.1 Geometrically Based Circular Macro-cell Model (GBCM) [63, 115, 200, 199]. This model is an extension of the CDSM presented in Figure 4.1 with the following characteristics

- The scatterers lie within a circular ring with radius R_D around the mobile.

- When a macro-cell is assumed, the antenna height is relatively large, resulting in no signal scatterers around the base station. In this case $R_D < R$ is assumed.

- The DOA, TOA, joint DOA and TOA distribution, Doppler shift and signal power information can be calculated from this model, making it very versatile. The joint DOA/TOA pdf at the base station is given by

$$p(\tau, \phi_r) = \begin{cases} \frac{(R^2-\tau^2c^2)(R^2c+\tau^2c^3-2\tau c^2 R\cos\phi_r)}{4\pi R_D^2 (R\cos\phi_r - \tau c)^3} & \forall \quad \frac{R^2-2\tau cR\cos\phi_r+\tau^2c^2}{\tau c - R\cos\phi_r} \leq 2R_D \\ 0 & \text{otherwise,} \end{cases} \quad (4.23)$$

while the joint DOA/TOA pdf at the mobile is given by

$$p(\tau, \phi_r) = \begin{cases} \frac{(R^2-\tau^2c^2)(R^2c+\tau^2c^3-2\tau c^2 R\cos\phi_r)}{4\pi R_D^2 (R\cos\phi_r - \tau c)^3} & \forall \quad \frac{R^2-\tau^2c^2}{R\cos\phi_r - \tau c} \leq 2R_D \\ 0 & \text{otherwise.} \end{cases} \quad (4.24)$$

4.2.5.2 Geometrically Based Elliptical (GBSBEM) Micro-cell Wideband Model [147, 148].

- This model is as shown in Figure 3.1 with joint DOA and TOA distribution given by (3.3).

- The model is intended for micro-cellular environments where the antenna heights are relatively low. In this environment it is assumed that multi-path scattering near the base station is just as likely as at the mobile.

- By choosing τ_m appropriately, the model will account for nearly all the power and DOA of the multi-path signals, since the longer delays will experience greater path loss, and hence have relatively low power compared to those with shorter delays.

- An important parameter in this model is the choice of τ_m since it determines both the delay spread and angle spread of the channel. Appropriate values for τ_m can be derived from Table 4.3 [64].

- A model similar to the GBSEM, is the elliptical subregions model. The main difference between the two models is in the selection of the number of scatterers and the distribution of the scatterers. In the ESM model, the ellipse of Figure 3.1 is first subdivided into a number of elliptical subregions. The number of scatterers within each region is then selected from a Poisson random variable, the mean of which is chosen to match the measured time delay profile data.

Criteria	Expression
Fixed maximum delay, τ_m	τ_m = constant
Fixed threshold L_T in dB	$\tau_m = \tau_0 10^{(L_T - L_s(r))/10 n_l}$
Fixed delay spread, σ_τ	$\tau_m = 3.244 \sigma_\tau + \tau_0$
Fixed maximum excess delay, τ_e	$\tau_m = \tau_0 + \tau_e$

Table 4.3. Criteria for selecting τ_m.

4.2.6 Gaussian Wide Sense Stationary Uncorrelated Scattering Model (GWSSUS) [330, 331, 332, 333]

This model is statistical in nature and makes assumptions about the form of the received signal vector in order to provide a general equation for the received signal correlation vector. A summary of the model follows.

- In this model, N_s scatterers are grouped into clusters in space in such a way that the delay differences within each cluster are not resolvable within the transmission bandwidth BW. Figure 4.11 shows the geometry corresponding to $d_{clus} = 3$, with mean DOA for the kth cluster denoted by ϕ_i. The model further assumes that each cluster remains constant over several data bursts, b.

- The model provides a fairly general result for the form of the covariance matrix.

- The received vector is assumed to be

$$\mathbf{r}_b(t) = \sum_{i=1}^{d_{clus}} \mathbf{v}_{i,b} s(t - \tau_i), \qquad (4.25)$$

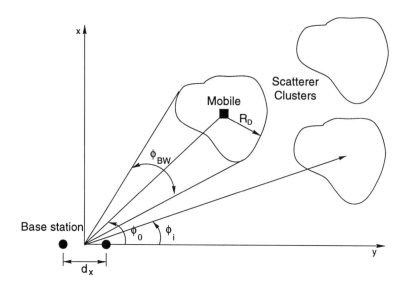

Figure 4.11. GWSSUS geometry.

where $\mathbf{v}_{i,b}$ is the superposition of the weight vectors during the bth data burst within the ith cluster, which can be expressed as

$$\mathbf{v}_{i,b} = \sum_{j=1}^{N_i} \beta_{i,j} \exp(j\varphi_{k,i})\mathbf{w}(\phi_i - \phi_{i,j}), \qquad (4.26)$$

where N_i denotes the number of scatterers in the ith cluster, $\beta_{i,j}$ the amplitude, $\varphi_{i,j}$ the phase, $\phi_{i,j}$ the DOA of the jth reflected scatterer of the ith cluster, and $\mathbf{w}$ the weight vector as defined in Chapter 2. It is further assumed that the steering vectors are independent.

- For $N_s > 10$, the elements of $\mathbf{v}_{i,b}$ can be assumed to be a multivariate, zero mean, complex Gaussian wide sense, stationary random process, with delays, τ_k assumed to be constant over several bursts. The phase $\varphi_{i,j}$ can change much more rapidly.

- The covariance matrix for the ith cluster is given by [332] as

$$\mathbf{R}_i = E\{\mathbf{v}_{i,b}\mathbf{v}_{i,b}^H\} = \sum_{j=1}^{N_i} |\beta_{i,j}|^2 E\{\mathbf{w}(\phi_i - \phi_{i,j})\mathbf{w}^H(\phi_i - \phi_{i,j})\}. \qquad (4.27)$$

- Frequency-selective fading channels can be modeled by including multiple clusters.

- The model reduces to the GAA [189] model when a single cluster ($d_{clus} = 1$) is considered. In this case the model is a narrow band channel model that is valid when the time spread is small compared to the inverse of the signal bandwidth, BW.

4.2.7 Raleigh's Model [331]

Raleigh's Model, also known as the time-varying vector channel model, has been developed to provide both small scale Rayleigh fading and theoretical spatial correlation properties. The model is compared to theoretical results in [214], with good agreement being achieved in regard to both time and spatial correlation properties. A short description of the model follows.

- The propagation environment is assumed to be densely populated with large dominant reflectors, as depicted in Figure 4.12. It is further assumed that the signal received by the base station is Rayleigh faded and that the angle spread is accounted for by dominant reflectors. The model therefore provides for both Rayleigh fading and theoretical spatial correlation.

- It is assumed that the channel is characterized by N_s dominant reflectors, which is a function of time.

- The received signal vector is given by

$$\mathbf{r}(t) = \sum_{i=0}^{N_s(t)-1} \mathbf{w}(\phi_i)\beta_i(t)s(t-\tau) + \mathbf{n}(t), \qquad (4.28)$$

where $\mathbf{w}$ is again the weight vector, $\beta_i(t)$ the complex amplitude, $s(t)$ the modulated signal, and $\mathbf{n}(t)$ additive noise.

- The complex path amplitude is given by

$$\beta_i(t) = E_{total_i}(t) \cdot \sqrt{\Gamma_i \mathbf{S}_M(\tau_i)}, \qquad (4.29)$$

where Γ_i accounts for log-normal fading, $S_M(\tau_i)$ describes the power delay profile and $E_{total_i}(t)$ is the complex radiation intensity profile as a function of time. The complex intensity is described by

$$E_{total_i}(t) = I \sum_{j=1}^{L} c_j(\phi_i) \exp(jf_d) \cos\phi_{r_j,i} t, \qquad (4.30)$$

where L is the number of signal components contributing to the ith dominant reflecting surface, I the antenna gains and transmit power. $c_j(\phi_i)$ is the complex radiation of the jth component of the ith dominant reflecting surface in the direction of ϕ_i. The maximum Doppler shift is f_d and $\phi_{r_{j,i}}$ the angle toward the jth component of the ith dominant reflector with respect to the motion of the mobile. The intensity, $E_{total_i}(t)$, exhibits a complex Gaussian distribution in all directions away from the mobile [214].

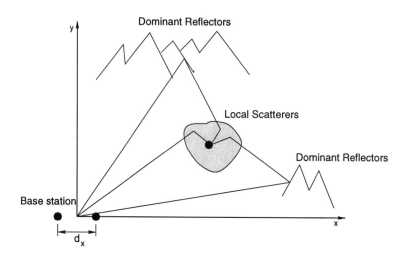

Figure 4.12. Signal environment for Raleigh's model.

4.2.8 Uniform Sectored Distribution (USD) Model [182]

- As depicted in Figure 4.13, this model assumes that all scatterers are uniformly distributed within an angle distribution of 2Δ and radial range ΔR centered around the mobile.

- The magnitude and phase of each scatterer is selected at random from a uniform distribution of $[0, 1]$ and $(0, 2\pi]$ respectively.

- As the number of scatterers approaches infinity, the signal envelope becomes Rayleigh faded with uniform phase.

- The model is useful to study the effect of angle spread on spatial diversity techniques.

- From this model it can be argued that beam steering techniques are most suitable for scatterer distributions with widths slightly wider than the beam widths.

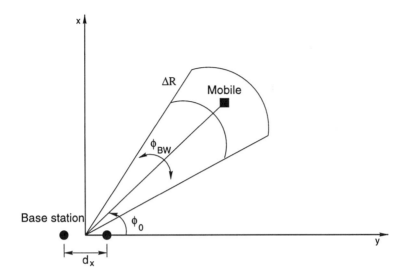

Figure 4.13. USD geometry.

4.2.9 Modified Saleh-Valenzuel's Model [244]

Based on experimental data, Saleh and Valenzuela developed a multi-path channel model for the indoor environment, based on the clustering phenomenon (which refers to the observation that multi-path components arrive at the receiver in groups or clusters) [231]. This model was modified by Spencer *et al.* [244] who were motivated by the need to include DOA information in the model.

- It is assumed that the time and the angle of arrival are statistically independent, i.e. $h(t, \phi) = h(t)h(\phi)$. The time and angular impulse responses are given by

$$h(t) = \sum_{i=0}^{\infty} \sum_{j=0}^{\infty} \beta_{i,j}^{(t)} \delta(t - T_{s_i} - \tau_{i,j}), \qquad (4.31)$$

and

$$h(\phi) = \sum_{i=0}^{\infty} \sum_{j=0}^{\infty} \beta_{i,j}^{(a)} \delta(\phi - \phi_i - \phi_{r_{j,i}}), \qquad (4.32)$$

where the sum over i corresponds to the clusters and the sum over j represents the rays within a cluster. Further, $\beta_{i,j}^{(t)}$ are Rayleigh distributed, $\beta_{i,j}^{(a)}$

is the amplitude of the jth ray in the ith cluster, ϕ_i is the mean angle of the ith cluster and is assumed to be uniformly distributed and $\phi_{r_{j,i}}$ corresponds to the ray angle within a cluster and is modeled as a Laplacian distributed random variable, with standard deviation σ_L, given by

$$p_{\Phi_r}(\phi_r) = \frac{1}{\sqrt{2}\sigma_L} \exp\left(-\left|\frac{\sqrt{2}\phi_r}{\sigma_L}\right|\right). \qquad (4.33)$$

4.2.10 Extended Tap-delay-line Model [129]

This model is an extension of the traditional statistical tap-delay-line model [205], which includes DOA information. The joint density functions of the model parameters should be determined from measurements as indicated in [129].

The channel impulse response can be written as

$$h(t,\tau,\phi) = \sum_{l=1}^{L} \beta_l(t)\delta(\tau - \tau_l)\delta(\phi - \phi_l), \qquad (4.34)$$

where L is the number of taps, associated with the time delay τ_l, the complex amplitude β_l and the DOA, ϕ_l.

4.2.11 Measurement-based Channel Models

One of the underlying difficulties with developing a propagation model for any environment is that no two areas are the same in the composition of the buildings and terrain. Therefore, it is often difficult to apply a model developed for one area to another area with a slightly different environment. Empirical models usually begin with a general qualitative classification into urban, suburban and rural areas with further quantitative characteristics, such as building density, area covered and locations, as well as the average building height, vegetation density and terrain variation being specified within each of these areas [192]. Okumura et al. [187] developed a series of curves used to predict the field strength of measurements in and around Tokyo for very high base station antennas at the VHF and UHF bands. In this approach the path loss between the base station and mobile in an urban area is found by adding free space loss to a median attenuation factor which is determined from empirical graphs at various frequencies and distances. From the original work by Okumura, Hata developed a series of computationally efficient formulas that calculated the propagation loss with small errors for frequencies in the range 100-1500 MHz. This formula is the standard for predicting path loss in urban areas and standard correction factors are added to account for large or medium-small cities and for suburban and rural areas.

Following on the results of various measurement campaigns such as those conducted by Ikegami [113, 114] and also the work of Walfisch [311], the European research committee, COST-231, developed a model that estimates the path loss over a series of buildings with uniform height and separation. The model is expected to provide an average error of around 4 dB and a standard deviation of 6 dB. The COST-231 Walfisch-Ikegami formulation has been compared with measurements in a number of European cities including Mannheim and Darmstadt in Germany [160] and Lisbon, Portugal [57] with slightly better than expected prediction results.

In addition to the above mentioned models, by characterizing the propagation environment in terms of scattering points, a channel model has been proposed in [32], that uses the measured data as parameters of the model. The time-variant channel impulse response takes the form

$$h_{tv}(t,\tau) = \int_0^{2\pi} h(t,\tau,\phi) E_{total}(\tau,\phi) f(\tau) d\phi \qquad (4.35)$$

where $f(\tau)$ is the impulse response representing the joint transfer characteristic of the transmission system components, such as the modulator, demodulator, filters etc. The function $E_{total}(\tau,\phi)$ is the base station antenna characteristic function. The time-variant directional distribution function of the channel impulse response, as seen from the base station, is represented by $h(t,\tau,\phi)$. Measurements are used to determine the distribution of $h(t,\tau,\phi)$, and are dependent on mobile velocity, location and orientation.

4.2.12 Ray Tracing Models

From the previous discussion of scattering models, it is clear that they are all based on statistical analysis and, therefore, give us information about the average path loss and average delay spread, according to the environment in which the base station - mobile system operates. A deterministic model, known as ray tracing has been proposed based on the geometric theory, reflection, diffraction and scattering models, i.e. [237, 229, 283, 241]. Realistic parameters for this model are acquired from site-specific information, usually available from building databases or geographic information. Although the ray tracing model can be more accurate than the statistical models, it carries a very high computational burden, and lack of geographic models makes it difficult to use.

The use of ray tracing models stems from the fact that standard measurement based models are generally only valid in areas with buildings of fairly uniform heights located on a regular street grid, and when the base station antenna is near or above rooftops [146]. For areas where the buildings are not uniform, or when the base station is immersed within the surrounding buildings, it is necessary to develop a site specific propagation model based on ray optics and the geometrical theory of diffraction [126]. These ray tracing models utilize the shoot and bounce ray method or the method of images and treat each building as an individual scatterer in order to predict the power at any given location. Reflections from the building walls are usually approximated by the Fresnel reflection coefficient [17], while diffraction from the building edges is approximated using the diffraction coefficient for a perfectly absorbing [72], perfectly conducting [134] or a dielectric wedge [165]. The use of ray tracing techniques for propagation prediction stems from indoor propagation modeling [109, 77], where the propagation environment is in a confined area and for the most part requires that only reflection and transmission at walls be accounted for.

A number of different ray tracing methods are currently used in propagation prediction. These include [146] two dimensional ray trace in the horizontal plane, methods employing 2D ray traces in orthogonal planes, ray tracing in full three dimensional space and the vertical plane launch (VPL) technique, which approximates a full 3D ray trace. When employed for micro-cells with ranges that are less than 1 kilometer, these techniques can provide accurate predictions of the propagation characteristics in a cluttered urban environment.

4.3 FADING DISTRIBUTION BASED ON SCATTERING ENVIRONMENT

One of the important requirements in a spatial-temporal channel model is to incorporate the effect of the non-homogeneous geography of the cell into the temporal fading model used. Consider the case where, for instance, each multi-path echo received at the base station is subject to Rayleigh fading. The question that arises is whether it is accurate to assume that all of the received multi-path echos will have the same statistical fading distribution. Is it not possible that some of the received paths may contain a LOS component changing the distribution of the fading envelope from Rayleigh to Rician, or perhaps that some paths may exhibit more severe fading, i.e. as described by a one-sided exponential distribution. Examining results from extensive measurements, this is the case. For example, in the experimental study in [262], the urban propagation channel is modeled as a Rician channel with *varying* $\mathcal{K}_k$ parameter. This indicates that an accurate model would describe the fading effects on each received multi-path signal at the base station.

Following the approach of [157, 158], the fading process on each of the received multi-path signals can be modeled by incorporating information on the DOA of the multi-path signals at the base station. Utilizing the properties of the Nakagami distribution, varying degrees of fading can be approximated by the correct choice of the m parameter. In order to develop a relationship between the fading exhibited by a signal and its DOA, the results of [262] were used in [158]. In all relevant cases considered (suburban worst-case, Typical Urban and Bad Urban), the cumulative distribution function (cdf) of the $\mathcal{K}_k$ parameter measured over the three strongest paths received is close to the cdf of a Gaussian pdf or two-sided exponential distribution [205]. Based on these results, two alternative models are used to create a relationship between the fading experienced by multi-path signals, embodied by the m-parameter and the DOA (ϕ). These models are described below.

4.3.1 Exponential Fading Distribution (EFD) [157, 158]

Firstly, a relationship between m and ϕ, based on a generalized exponential function, is established by setting m in (5.21) as equal to

$$m(\phi) = m_0 e^{-\delta_m(|\phi_0 - \phi|)} \qquad m(\phi) > 0.5 \qquad (4.36)$$

where m_0 denotes the Nakagami parameter of the main received path, ϕ_0 the DOA of the main received signal path and δ_m a parameter to control the increase in the severity of the fading as a function of the angular spread of the multi-path signals arriving at the base station. When $\delta_m = 0$, this model represents a standard, constant m Nakagami fading channel. The changes in m as a function of ϕ (that is the value of δ_m) will depend on the local scattering environment model. Clearly δ_m is dependent on the physical environment in which the mobile system is to operate and values for δ_m should be determined through measurement. However, as with the rationale that was used in the development of the local scattering models, the EFD model would simulate a situation where signals received directly from the mobile (e.g. LOS components) exhibit less fading than signals with DOA's offset from the DOA of the mobile. To use this model, an accurate estimation of δ_m is needed which is dependent on the type of signalling environment. Based on arguments from [262], Figure 4.14 and Table 4.4 give typical values for δ_m and the Nakagami-m fading parameter.

4.3.2 Gaussian Fading Distribution (GFD) [157, 158]

Alternatively, based on the fading envelope distribution measurements in [262], the relationship between m and ϕ can be written as

$$m(\phi) = m_0 e^{-\frac{(\phi_0 - \phi)^2}{2\sigma_m^2}} \qquad m(\phi) > 0.5. \qquad (4.37)$$

In (4.37), the fading parameters of the multi-path components as a function of ϕ are controlled by σ_m, the variance of the fading parameter. Setting σ_m equal to ∞ will yield a constant m Nakagami fading channel. Decreasing σ_m will yield non-constant fading distributions with the fading effects becoming increasingly severe as the DOA of the multi-path signal is removed from the DOA of the main received signal path. Figure 4.14 and Table 4.4 give typical values for σ_m and the Nakagami-m fading parameter.

4.4 SUMMARY OF SPACE-TIME CHANNEL MODELS

Thus far, all the elements required to accurately model the cellular channel have been presented. In Chapter 3, the concepts of fading, scattering environments, subscriber distribution and correlation were introduced. These aspects were further developed in this chapter where more comprehensive treatments of the CDSM and GS model were presented. In addition, the distribution of the fading

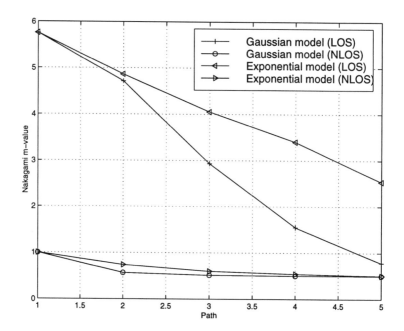

Figure 4.14. Fading parameters of multi-path echoes received at the base station ($\delta_m = 1, \sigma_m = 0.6$).

Environment	Temporal Fading Model	m_0	Fading Decay
Bad Urban (NLOS)	Gaussian	1	$\sigma_m = 0.6$
Typical Urban (LOS)	Exponential	≈ 6	$\delta_m = 1$
Macro-cell	Exponential	> 6	$\delta_m < 1$

Table 4.4. Summary of temporal fading model.

envelope of multi-path signals was explained in Section 4.3. In this section, all of the aspects of the channel model described in detail in the preceding section are pulled together and it is shown how a comprehensive channel model for use in the evaluation of cellular system models can be constructed.

Figure 4.15 depicts the process of constructing a channel model as a simple flow diagram. Let us first turn our attention to the choice of a cellular environment. This choice between the different cellular environments can firstly be viewed as a choice between a high-rank and a low-rank channel model. Environments with very low angular spread of the received signals are deemed to be low-rank channels and, in these cases, the description of the local scattering environment becomes less important. On the other hand, when the angular

SPACE-TIME CHANNEL MODELS   89

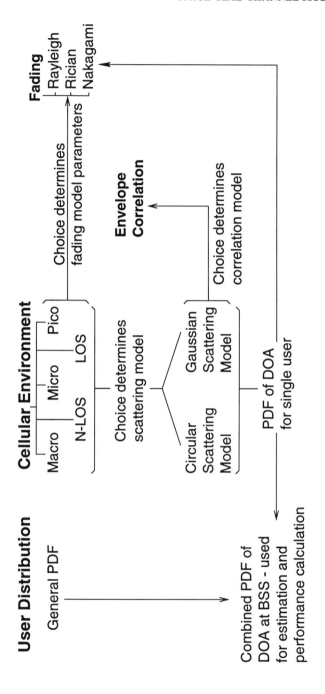

Figure 4.15.   Constructing a typical cellular channel model.

spread of the received signals is expected to be larger, the channel model can be described as high-rank, and consequently the description of the local scattering environment is of higher importance. Thus, in the case of a high rank channel model (which would be the predominant case in cellular environments), a decision must be made as to whether to describe the scattering area surrounding the base station or a mobile using the CDSM or the GS model. Both models will yield accurate results, with the GS model yielding a more general description with average parameters defined in Table 4.1.

The choice of the scatterer model will determine three additional channel model parameters namely the correlation between the received signal envelopes at different points in space, the pdf of the received signal at the base station and the characteristics of the temporal fading present on each multi-path signal. The correlation between the received signal envelopes at different points in space will determine the possible gains that can be achieved using diversity systems such as those described in Section 5.2.2. As the possible diversity gain is significantly influenced by the correlation between the fading envelopes, this part of the channel model is extremely important in order to accurately estimate the overall system performance.

Whereas the correlation characteristics of the channel model influence mainly the diversity performance of the system, the pdf of the DOA at the base station influences the performance of a beamforming system, as well as the fading characteristics of the received signals (see below). The specific characteristics of the pdf will be determined by the scattering model chosen and again the Gaussian model with the average parameters presented in Table 4.1 gives average performance results for typical micro-, macro- and pico cellular environments.

Finally, the choice of cellular environment will determine the characteristics of the fading envelope of each received multi-path signal. Generally, the received signal envelope is described as either Rayleigh, Rician or Nakagami. The Nakagami description provides the most general description of the fading characteristics and the use of this distribution is described in Section 4.3. In order to accurately determine the fading characteristics of each received multi-path signal, the pdf of the DOA of signals at the base station is also required (see Section 4.3). Typical values used to describe the fading characteristic are shown in Table 4.4.

In addition to the choice of the environment, the choice of the user distribution will also significantly influence the performance of the cellular system. Thus, an estimation of the distribution of the user population based on the geographical environment where the users are active is required as one of the core ingredients of the channel model, specifically in the case of systems including beamforming. The general assumption of a uniform user distribution will tend to yield average performance results. Once the pdf of the DOA of a single user and the distribution of all users are known, it is easy to determine the combined pdf of the DOA of signals at the base station.

With the process described in Figure 4.15 completed, a comprehensive description of the cellular environment is available. This description incorporates the major aspects that influence the performance of a cellular system and therefore the resulting channel model can be used to evaluate the performance of any cellular system under a variety of conditions. In the following chapters, the specific analysis techniques required to determine the performance of cellular CDMA systems incorporating smart antenna techniques will be presented.

4.5 SUMMARY

This chapter provided a detailed overview of channel models for space-time processing. Specifically correlation effects were discussed due to its importance in beamforming and diversity systems. A model for temporal fading was also discussed to model the fading process of the received multi-path components more realistically. Key issues that impact on the performance of specifically WCDMA were also presented.

5 SMART ANTENNA TECHNIQUES

We have seen that factors such as the scattering environment, the user distribution and the fading environment, amongst others, play important roles in determining the performance of space-time mobile systems. In this chapter four smart antenna techniques, beamforming, diversity, sectorization and switched-beam systems are considered in more detail to mitigate the effects due to the inherent impairments of the mobile communication channel. These techniques are especially applicable to mobile CDMA communications, where interference is more pronounced due to the fact that users transmit on the same frequency with unique spreading codes.

The aim of a well-designed mobile communication system is to share the common transmission medium in such a manner that

- the average overall amount of transmitted information is as large as possible,
- the average probability of error experienced by each user is as low as possible, and
- the average delay is as low as possible.

In general, not all of these goals can be achieved simultaneously, and the design process involves a trade-off between these aims. By making use of coding, multiuser detection and smart antenna techniques, it is possible to approach these goals. For instance, beamforming can be used to decrease a system's probability of error by reducing CDMA interference. This can be achieved by intelligent combination of the received signals by multiple antenna elements at the base station or mobile. In a mobile communication system with antenna arrays, the fast fading signal component introduces a random phase and amplitude to the received signal on each antenna element, which perturbs the steering vector of the array. In the case of Rayleigh or Nakagami fading, the phase can take on any value between $(0, 2\pi]$, and the DOA of the waves may be impossible to determine from short-duration observations of the received signal. Similarly, the concept of an array radiation pattern relies on the plane wave which is incident on the array elements to have constant amplitude. In a fading environment it may therefore not be useful to implement the beamformer to create lobes and nulls toward desired and interfering sources. When the fast fading is highly correlated between the elements, it may be considered as a single scalar which multiplies with the steering vector, affecting all elements equally. It may then be possible to recover the steering vector with techniques discussed in Section 5.1.3. On the other hand, no receive diversity gain can be obtained, since receive diversity relies on uncorrelated fading. The correlation between elements decreases with element spacing and changes according to the scattering environment in which the system operates. There is therefore a clear conflict between the avoidance of grating lobes and the need for receive diversity gain. In general, larger angle spreads and element spacing result in lower correlations, which provide an increased receive diversity gain for fixed element spacing. Measurements of the correlation at both the base station and mobile are consistent with a narrow angular spread at the base station and a large angular spread at the mobile. Correlation measurements at the base station have shown that the typical radius of scatterers is from 100 to 200 λ [142]. This explains how a narrowband signal at a mobile receiver only requires about 0.2λ spacing for good receive diversity (i.e. low correlation between elements), whereas comparably good receive diversity at a tall base station requires a spacing of about 40λ. The influence of correlation on the beam pattern is discussed in Section 5.1.4.

In the absence of the concept of radiation patterns, the significant measure of "distance" between sources is actually the correlation of their propagation channels. Two mobiles may be effectively separated in SDMA provided they are essentially uncorrelated. In a classical Rayleigh channel, the necessary distance between two mobiles is then of the order of $\lambda/2$, where λ is the wavelength. When the M_D elements of the antenna array have completely uncorrelated fading and no interferers are present, the optimum combiner performance is identical to M_D-branch maximum ratio combining (MRC) receive diversity. When an interferer is introduced, the receive diversity order is reduced by one because one degree of freedom is lost in nulling the interferer.

From the discussion above it is evident that mitigation of channel effects is possible by better receiver design. However, due to the strict complexity requirements of terminals and the characteristics of the downlink, a brute-force application of advanced detectors with multiple receive antennas is not a desired solution to the downlink capacity problem. Multiple transmit antennas

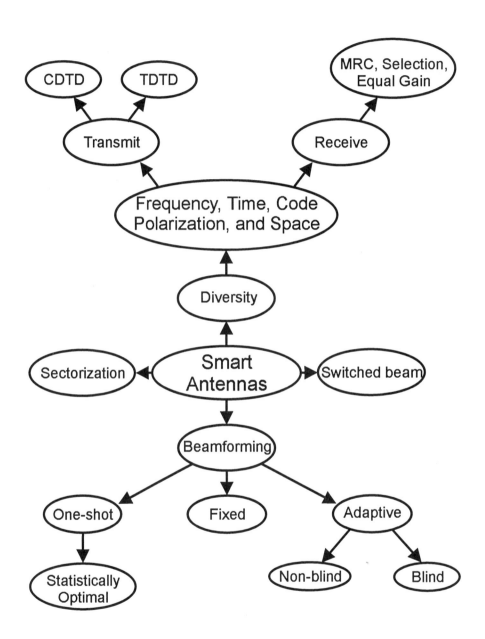

Figure 5.1. Classification of smart antenna techniques.

(or transmit diversity) at the base station will increase the downlink capacity with a marginal increase in mobile terminal complexity.

Smart antenna techniques can be classified according to Figure 5.1. Each of these components are described in the remainder of this chapter.

5.1 BEAMFORMING

As the DOA of the received signals at the base station are time-varying due to movement of the mobile and also due to the scattering environment, the co-channel interference, multi-path components and even Doppler frequency are time varying functions. By using a beamformer, it is possible to separate signals collocated in frequency, but separated in the spatial domain and to track these time varying signals. In a CDMA system, this specifically results in reducing the interference from unwanted signals by optimizing the array pattern through the adjustment of the weights of the array according to some criteria, or cost function.

From Chapter 2 where the general principles of antenna arrays, as well as the general structure of a space-time processor were described, it follows readily that the received signal in a coherent multiuser CDMA system is (see (2.14))

$$\mathbf{r}(t) = \sum_{k=1}^{K} \sum_{l=1}^{L} A^{(k)} b^{(k)}(t) a^{(k)}(t) \hat{c}_l^{(k)}(t) e^{-j(\omega_c t + \theta^{(k)})} \mathbf{s}_a(\phi_l^{(k)}) + \mathbf{n}(t), \qquad (5.1)$$

with the output of the array as defined in (2.7). The task of a beamforming system is to find a $\mathbf{w}$, or set of $\mathbf{w}$'s that will result in the antenna array beam pointing in the desired direction or directions. The simplest case will be when the pointing direction(s) do not change, and is known as a fixed beamforming array.

5.1.1 Fixed Beamforming

The first application of antenna arrays in beamforming is that of fixed beamforming networks. Assume for instance that M_B beams are to be formed using an M_B element ULA. Such a scenario could be required when switched-beam (see Section 5.4) or sectorization (see Section 5.3) schemes are to be implemented. As an example consider a 6-sector cell. Generally, the sectors will be formed using six, 60° directional antennas. Alternatively, a six element beamforming array can be used with six different weight vectors, each forming a 60° beam in a different direction.

In order to construct such a system, an algorithm to determine the weight vector for a certain pointing direction is required. Such an algorithm will normally be known as an array synthesis algorithm. Array synthesis algorithms often use orthogonal polynomials, such as Chebyschev polynomials, to synthesize a beam pattern. These algorithms are thoroughly discussed in [16] and will not be treated in detail as they are special cases of adaptive beamforming algorithms which are more suitable for mobile communications.

5.1.2 One-shot Beamforming

A slightly more general case of a fixed beamforming network would be a one-shot beamformer. We define a one-shot beamformer as a beamforming array where an optimal radiation pattern or antenna weights are determined using a single operation. Such beamforming techniques are also known as statistically optimal techniques as they determine a weight vector which is optimum in some statistical sense. Specifically, the weight vector is determined by minimizing a cost function. Minimizing the cost function will maximize the signal quality at the output of the beamformer. One of the most popular techniques is the minimum mean square error (MMSE) algorithm, which is widely used. Whereas we will use this technique to determine a statistically optimal antenna weight vector, the same algorithm is used in, for example, multiuser detection algorithms (see Section 7.3.3.2).

In a beamforming environment, the MMSE algorithm will minimize the difference between the array output $\mathbf{r}_o(t) = \mathbf{w}_j^H \mathbf{r}(t)$ and $d^{(j)}(t)$, a locally generated replica of the desired signal of the jth subscriber, by finding a suitable weight vector $\mathbf{w}_j$. Specifically, the MMSE approach will minimize the cost function

$$J(\mathbf{w}_j) = E\left\{ \|\mathbf{w}_j^H \mathbf{r}_{o,i} - d_i^{(j)}\|^2 \right\}, \tag{5.2}$$

where $\mathbf{r}_{o,i} = \mathbf{r}_o(iT_{samp})$ and $d_i^{(j)} = d^{(j)}(iT_{samp})$ with T_{samp} denoting the sampling rate [149]. This cost function is the expected value of the square error between the array output for the jth signal and the desired version of that signal at time index i. Rewriting (5.2) as

$$\begin{aligned} J(\mathbf{w}_j) &= \mathbf{w}_j^H E\left\{\mathbf{r}_{o,i}\mathbf{r}_{o,i}^H\right\} \mathbf{w}_j - E\left\{d_i^{(j)}\mathbf{r}_{o,i}^H\right\} \mathbf{w}_j \\ &\quad - \mathbf{w}_j^H E\left\{\mathbf{r}_{o,i}d_i^{(j)*}\right\} + E\left\{d_i^{(j)}d_i^{(j)*}\right\}, \end{aligned} \tag{5.3}$$

where $()^*$ indicates complex conjugate.

This function can be minimized by determining the location where the gradient equals zero. Following the method in [104], the gradient of the cost function can be written as

$$\nabla J(\mathbf{w}_j) = 2\mathbf{R}\mathbf{w}_j - 2\mathbf{p}, \tag{5.4}$$

where $\mathbf{R}$ is the correlation matrix of the data vector,

$$\mathbf{R} = E\left\{\mathbf{r}_{o,i}\mathbf{r}_{o,i}^H\right\}, \tag{5.5}$$

and $\mathbf{p}$ is the cross-correlation between the data vector and the desired signal,

$$\mathbf{p} = E\left\{\mathbf{r}_{o,i}d_i^{(j)*}\right\}. \tag{5.6}$$

To find the optimum weight vector, (5.4) is set equal to zero yielding a weight vector of

$$\mathbf{w}_j = \mathbf{R}^{-1}\mathbf{p}. \qquad (5.7)$$

The solution in (5.7), requires the inversion of the correlation matrix. This implies that the correlation matrix should be non-singular in order to yield a solution and also that the desired signal should be known. For the MMSE algorithm, this can be accomplished by periodically sending a specific training sequence to the receiver. Thus, a new weight vector can be determined each time a new training sequence is received. However, each time a training sequence is received, (5.7) must be solved. Alternatively, (5.7) can be solved iteratively yielding an adaptive beamforming algorithm.

5.1.3 Adaptive Beamforming Algorithms

There are several reasons why it is not desirable to solve (5.7) directly. Since the mobile environment varies with time, the solution of the weight vector must be updated periodically. Typically, the change in the channel from one adaptation cycle to the next will be small. Also, since the data required to estimate the weight vector is noisy, it is desirable to use the current weight vector to determine the next weight vector. This would result in a smoothing of the weight vector reducing the effect of noise. When a training sequence is used in the adaptation process, the beamforming method is known as non-blind beamforming, while the solution of (5.7), without the use of training sequences, is known as blind beamforming.

5.1.3.1 Non-blind beamforming. An iterative solution of (5.7) can be written as

$$\mathbf{w}_{j,i+1} = \mathbf{w}_{j,i} - \frac{1}{2}\mu \nabla J(\mathbf{w}_{j,i}), \qquad (5.8)$$

where μ is the convergence factor which controls the rate of adaptation [149]. This recursive solution of finding the minimum of the cost function is known as a stochastic gradient approach. This function is parabolic, which means that it has an unique minimum value [104]. Two of the most common algorithms that can be used to solve (5.8) are the least mean square (LMS) and recursive least squares (RLS) algorithms.

In the case of the LMS algorithm, (5.8) can be written as

$$\mathbf{w}_{j,i+1} = \mathbf{w}_{j,i} + \mu \mathbf{r}_{o,i} e_i^*, \qquad (5.9)$$

where

$$e_i^* = d_{j,i} - z_i, \qquad (5.10)$$

and
$$z_i = \mathbf{w}_{j,i}^H \mathbf{r}_{o,i}. \quad (5.11)$$

Furthermore,
$$\mathbf{w}_{j,0} = 0, \quad (5.12)$$

and the convergence coefficient is in the range
$$0 < \mu < \mathrm{tr}(\mathbf{R}). \quad (5.13)$$

In the adaptive algorithms that solve (5.7), the desired signal can either be a training sequence or it can be generated using decision directed learning. When a training sequence is used, the adaptation of $\mathbf{w}_j$ is performed every time a new training sequence is received. It is therefore assumed that the channel is stationary during the interval between training sequences. Furthermore, the overall system capacity is reduced by the extra capacity that is used by the training sequence. When decision directed learning is used, the receiver uses recreated modulated symbols based on symbol decisions as the desired signal to adapt the weight vector. The disadvantage of this approach is that errors in the decision directed process can lead to poorly directed weight updates which in turn lead to increased decision errors. Typically, the decision directed approach does not work well in environments with low SNR. However, the approach of using decision directed learning opens the opportunity to integrate the processes of beamforming, multiuser detection and space-time coding as the output of the multiuser detector or space-time coder can be used as the reference signal for the beamforming algorithm.

5.1.3.2 Blind Beamforming. Blind beamforming techniques determine optimum weight vectors without the use of training signals. Blind beamforming in CDMA systems do not use training sequences but in general do assume that the spreading codes are known. Although this is the general approach, several algorithms have recently been proposed which assumes no prior knowledge of the spreading codes [275, 73].

The objective of blind beamforming algorithms is similar to non-blind techniques, that is, to determine a set of weights which will minimize some cost function in order to maximize the quality of the received signal. Whereas non-blind algorithms utilize a copy of the desired signal to perform the minimization, blind algorithms exploit the temporal, spatial or code structures of the signal such as non-Gaussianity, a constant modulus property, cyclostationarity or the array manifold to perform the same task.

5.1.3.3 Constant Modulus Algorithm (CMA). Let us consider the CMA algorithm as an example of a blind beamforming algorithm. The CMA algorithm may be applied to signals with constant envelopes yielding a cost function

$$J(\mathbf{w}_j) = E\left\{|||\mathbf{w}_j^H\mathbf{r}_{o,i}||^p - ||\alpha_{cma}||^p|^q\right\}, \tag{5.14}$$

where α_{cma} is the desired signal amplitude at the array output. An adaptive array using the CMA will attempt to ensure that the array yields a constant amplitude output. The exponents p and q are each equal to either 1 or 2, with different values yielding different steepest descent algorithms with varying convergence characteristics and complexity.

For $p = 1$, $q = 2$ and $\alpha_{cma} = 1$, the weight vector obtained from the CMA algorithm is given by [149]

$$\mathbf{w}_{j,i+1} = \mathbf{w}_{j,i} - \mu \mathbf{r}_{o,i} e_i^*, \tag{5.15}$$

where

$$e_i = 2\left(z_i - \frac{z_i}{||z_i||}\right), \tag{5.16}$$

and

$$z_i = \mathbf{w}_{j,i}^H \mathbf{r}_{o,i}. \tag{5.17}$$

Clearly, no estimate of the desired signal is required since the new weight vector depends only the array output and the previous weight vector.

An alternative method to minimize the constant modulus cost function is to use a least squares (LS) approach. This method was proposed by Agee [4] and forms the basis of a number of integrated beamforming approaches that will be treated later. Specifically, for the LS-CMA, the weight vector can be written as

$$\mathbf{w}_{j,i+1} = \left[\mathbf{R}_o \mathbf{R}_o^H\right]^{-1} \mathbf{R}_o \mathbf{e}^*, \tag{5.18}$$

where $\mathbf{R}_o = [\mathbf{r}_o(1)\ \mathbf{r}_o(2) \cdots \mathbf{r}_o(K_a)]$,

$$\mathbf{e} = \left[\frac{z_1}{||z_1||}\ \frac{z_2}{||z_2||}\ \cdots\ \frac{z_{K_a}}{||z_{K_a}||}\right], \tag{5.19}$$

and K_a is the length of a block of observations.

Since all users in a CDMA system are being received simultaneously, a disadvantage of the CMA is that the algorithm will simply lock onto the strongest signal, which may be an interfering signal. Therefore, in a CDMA environment, the blind beamformer must use additional information about the wanted signal, such as the spreading code, to ensure that the beam is optimized for the desired user. Specifically, blind beamforming algorithms for CDMA systems can use the following sets of additional information to aid the beamforming algorithm

- knowledge of the DOA of desired and interfering signals,

- structural properties of the signal such as the CM, and

- decisions made on previous symbols.

In the following sections, some of the more well known blind algorithms are discussed.

5.1.3.4 Algorithms based on DOA Estimation. Well known high-resolution techniques for DOA estimation include the MUSIC [240, 239] and ESPRIT [195] algorithms. After DOA estimation, the optimum beamformer is constructed from the corresponding array response. Drawbacks of these techniques are that the performance strongly depends on

- the reliability of prior spatial information, (such as the array manifold) and in many practical situations the spatial information is not available or unreliable, and

- the high computational complexity. The number of DOA's that these algorithms can estimate is limited by the number of array elements (which has very limited use in a multi-path CDMA system where the number of users' signals far exceed the number of antenna elements) and these algorithms do not exploit the known temporal structure of the received signal.

5.1.3.5 Algorithms based on Property-restoral Techniques. Since communication signals in general, and specifically digital communications signals, possess unique properties, such as constant modulus or spectral self-coherence properties, the adaptive algorithm can try to restore these properties after the signal has been corrupted by noise and fading through the mobile channel. By restoring these properties, the output of the array will be an estimated reconstructed version of the transmitted signal. A well known and frequently used algorithm that uses the constant modulus nature of the signal is the CMA [184, 273, 274] described in Section 5.1.3.3. Other variants of the CMA (with processing in the data space) include the

- least-squares CMA (LS-CMA) [4],

- linearly constrained CMA [230] (which transforms *a priori* knowledge of the signal into a set of linear constraints on the weight vector),

- Marquardt method [84] (which uses the method of non-linear least squares and steepest-descent to update the weights),

- multistage CMA beamformer [248, 171, 172] (which is comprised of a cascade of constant modulus array subsections, each capturing one of the signals impinging on the array, followed by an adaptive canceler each to remove the captured signals from the input before processing by subsequent sections),

- orthogonalized CMA (O-CMA) [92] (which orthogonalizes the correlation matrix of the received data by multiplying the scaled instantaneous estimate of the gradient vector of the original CMA by the inverse of the correlation matrix of the input data),

- recursive CMA (R-CMA) [201] (which is related to the RLS non-blind algorithm, and uses an exponentially weighed squared error as a cost function over a time window), and the

- iterative least squares with projection CMA (ILSP-CMA) [191] (which estimates the steering vector by using the iterative least squares with projection method).

Variants of the CMA that perform adaptation in the beam space include the beam space CMA (BS-CMA) [40, 41], where a multi-beam is formed by means of a FFT and only the beams whose output power exceeds a threshold level are selected. It is also claimed in [41] that the BS-CMA converges faster than the element space CMA (EL-CMA). Digital communications signals also exhibit cyclostationary properties which can be used for beamforming purposes. Specifically, the periodicities in the second order statistics of a signal lead to the existence of correlation between the signal and frequency shifted versions of itself for certain discrete frequency shifts, and is known as spectral self-coherence. Based on this property a class of spectral self-coherence restoral (SCORE) algorithms were developed in [5], of which the least squares SCORE, the cross-SCORE and the auto-SCORE are the most well-known. Modifications of the SCORE algorithms were also proposed in [31] to increase convergence speed and reduce computational complexity. In a CDMA environment, these algorithms are of limited use, since they cannot separate multiple signals having the same cyclic features.

5.1.3.6 Algorithms based on Discrete Signal Structure. These algorithms exploit the discrete-alphabet property of digitally modulated communication signals. In [264, 263] the iterative least squares with projection (ILSP), the iterative least squares with enumeration (ILSE) and the recursive least squares enumeration (RLSE) algorithms are proposed and analyzed. The limitation of these algorithms is that they assume that the signals arrive synchronously at the array, which is not necessarily true in a mobile communication environment.

5.1.3.7 General Algorithms. In addition to more general beamforming algorithms proposed in [131, 120], the following algorithms are reported in the literature.

2D RAKE receiver [256, 128].
It is well known that a RAKE receiver combines multi-path components of a signal to improve system performance [205] and is used extensively in CDMA systems. A problem arises when the delay between multi-path components

SMART ANTENNA TECHNIQUES   103

is less than the chip period of the spreading sequence, and the multi-path components cannot be resolved. This problem can be overcome by the 2D RAKE receiver, where the spatial property of each multi-path component is exploited in such a way, that the components with less than a chip period delay, but with different DOA's, can still be resolved and optimally combined. This receiver estimates the pre-processing (before matched filtering) and post-processing array correlation matrices. The steering vector and noise covariance matrix are then used to estimate the beamforming weight vectors and try to steer a beam to each multi-path component.

Decision Directed (DD) Algorithms [164, 106].
In this algorithm, the beamformer attempts to adapt by using the demodulated signal as reference to update the beamforming weight vectors. An obvious limitation to this algorithm is that decisions made by the receiver will only be accurate at low average error rates, or relatively high SNR levels. When the error rate is indeed small, the adaptive beamformer is able to improve the weight vector by virtue of the correlation procedure built into its feedback control loop. An algorithm was proposed in [106] to allow soft transition between the CMA and the DD algorithm for PSK modulation. This algorithm was shown to have the robustness of the CMA at the beginning of the convergence phase, and the precision of the DD behavior near the convergence point.

5.1.3.8 Integrated Algorithms for CDMA Signalling. As mentioned before, a beamformer that will work effectively in a CDMA environment, where multiple users occupy the same frequency band, should attempt to form a beam directed to each user in such a way as to reduce interference from other directions. The beamformer should therefore generate K sets of complex weights (where K is the number of users in the system), corresponding to K different beam patterns, used to linearly combine the input data array vector to generate K different output ports. A structure known as a multi-target-type blind algorithm can be used to achieve this criteria, as proposed in [226]. Four specific algorithms for CDMA signalling have been proposed:

Multi-target Least Squares CMA (MT-LSCMA) [4].
This algorithm is computationally complex and contains three principal components: a soft-orthogonalized dynamic LS-CMA, a set of sorting and classification algorithms, and a fast acquisition algorithm. Since the LS-CMA utilizes only the *a priori* information that the received signals have a constant envelope, the weight vectors in different ports may converge to the same pattern. One of the steps to prevent this from happening is to perform Gram-Schmidt orthogonalization [205].

Multi-target Decision Directed (MT-DD) [226, 228].
This algorithm is equivalent to the MT-LSCMA with a DD algorithm replacing the LS-CMA algorithm. The disadvantage of this algorithm, relative to the MT-LSCMA, is that the signal constellation is not restricted to the

unit circle, but rather to ±1, resulting in a phase ambiguity. Depending on the type of DD algorithm used, i.e. least squares DD (LS-DD) or the steepest decent DD (SD-DD), there is a trade-off between convergence speed and computational complexity when compared to the MT-LSCMA.

Least Squares Despread Respread Multi-target Array (LS-DRMTA) [226, 228].

The previous two algorithms do not utilize the spreading code of the CDMA users, whereas the LS-DRMTA (and the LS-DRMTCMA) algorithm does. Computational complexity, relative to the previous algorithms, is reduced since orthogonalization and sorting is not needed to extract each different users' signal. A very important advantage of this algorithm in CDMA is that the number of output ports of the beamformer is not limited to the number of antenna elements of the array. If, in the MT-LSCMA and MT-DD algorithms, the number of users in the system is greater than the number of antenna elements, several users must share the same output port, and the interference cannot be reduced to a very low level. In this algorithm, since no orthogonalization is needed, the number of output ports can equal the number of users, even if the number of users is greater than the number of antenna elements. A drawback of this algorithm is its sensitivity to the initial data bit estimation which can result in a phase ambiguity of π for BPSK.

Least Squares Despread Respread Multi-target CMA (LS-DRMTCMA) [226, 228, 227].

In this algorithm the properties of the LS-DRMTA and the MT-LSCMA algorithms are combined to utilize both the spreading sequence information and the constant modulus property of the transmitted signal to adopt the weight vector. This algorithm has the advantages of the LS-DRMTA, with the additional advantage that it can achieve much lower error rates than the LS-DRMTA. This performance increase comes at the expense of increased complexity.

Figures 5.2 to 5.4 show the spatial filtering performance of a $M_B = 8$ ULA using the LS-DRMTCMA algorithm when three CDMA users, situated respectively at 60°, 0° and −60°, are operating in an AWGN environment. From these figures we see that the user of interrest's signal can be extracted by nulling out the interfering users. Due to the existence of noise, the nulls are not constructed perfectly, but the desired user has a gain of 15-20 dB above the interfering users. Due to the limitations in the scan angle of a ULA, users situated near the end-fire of the array, will be more difficult to isolate. To overcome this problem, a circular array will be more suitable in cellular applications. Section 5.1.4 describes the scan angle of an antenna array in more detail.

Figure 5.5 depicts the performance of the LS-DRMTCMA algorithm operating in a two-path Rayleigh fading channel. This figure illustrates how an antenna array can be used to reduce inter-symbol-interference. The array extracts the path with the strongest power (in this case at −10°) and reject the other path (which is located at 0°).

SMART ANTENNA TECHNIQUES   105

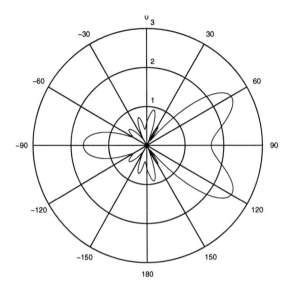

Figure 5.2. LS-DRMTCMA ULA beampattern for reference user located at $60°$ and interfering users at $0°$ and $-60°$ in AWGN.

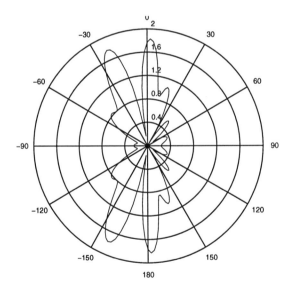

Figure 5.3. LS-DRMTCMA ULA beampattern for reference user located at $0°$ and interfering users at $60°$ and $-60°$ in AWGN.

106   SPACE-TIME PROCESSING FOR CDMA

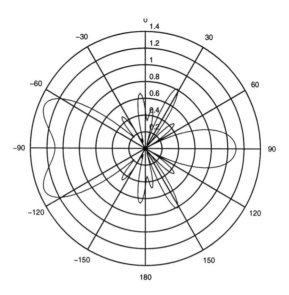

Figure 5.4. LS-DRMTCMA ULA beampattern for reference user located at $-60°$ and interfering users at $0°$ and $60°$ in AWGN.

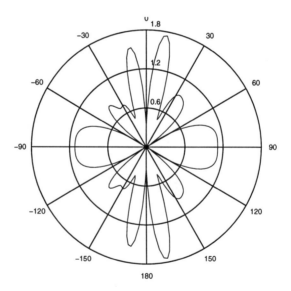

Figure 5.5. LS-DRMTCMA ULA beampattern in two-path Rayleigh with reference user located at $-10°$ and multi-path signal at $0°$.

5.1.4 Correlation Influence on Beam Pattern

As mentioned earlier, lack of correlation influences the beamforming capabilities of an antenna array, but increases the receive diversity gain. In this section we consider, by means of examples, the influence of correlation on the beam pattern. Figures 5.6 and 5.7 show the beam pattern for a $M_B = 12$ element ULA with antenna element spacing of $\lambda/2$. For each correlation value, 1000 snapshots were simulated and the corresponding radiation patterns generated. To establish a measure of degradation for the beamformer, these 1000 beams were plotted and the envelope extracted for each correlation value. Shown in these figures are the maximum beam pattern, the minimum beam pattern and the average beam pattern. Correlation of $\rho_{ij} = 0.8$ already degrades the beamforming characteristics and the beam pattern tends towards an omni-directional antenna pattern. However, the main beam is still at 0° with the side lobe levels altered. For correlation between the antenna elements of $\rho_{ij} = 0.3$, it is clear that the antenna array cannot be used as a beamformer - the antenna pattern tends toward an omni-directional pattern, with very little gain relative to the side lobes in the desired direction of 0°.

The use of an ULA as beamformer in a cellular system, especially in adaptive antenna array applications such as SDMA, is limited due to the scanning characteristics of the ULA geometry. A much better geometry for cellular use is a circular array. Consider for instance the $M_B = 12$ element circular array of Figure 5.8. Radiation patterns in the horizontal plane for scan angles $\phi_s = \{0°, 5°, 10°, 15°\}$ are shown in Figure 5.9 for the $M_B = 12$ element circular array with antenna spacing of $\lambda/2$. It is clear that the main beam stays virtually constant as a function of scan angle, with slight variations in the side lobe distributions. This beam pattern corresponds to the two extreme cases where the main beam is either directed towards one of the array elements, or where the main beam is directed between two of the radiating elements, it is clear that the array can be scanned through the full 360°, while still keeping the side lobe level within acceptable limits (in this case lower than 15 dB).

From Chapter 4 we know that the envelope correlation in a space-time system is directly proportional to the scattering environment. To show the effect of correlation on the beam pattern of the circular array of Figure 5.8, three channels models are considered: NLOS, LOS, and macro-cells. For these models the correlation, as a function of distance between elements, was determined from the GS model, described in Section 4.2.2. With reference to Figure 5.8, the correlation values for the three models at the different element locations are indicated in Table 5.1. Again 1000 beams are used to extract the envelope as indicated in Figures 5.10, 5.11 and 5.12. For these figures, it is assumed that elements $M_B = 1, 3$, and $M_B = 10, 12$ of the circular array will not contribute to the radiation pattern at $\phi = 0°$ due to the directive nature of the elements. It is clear that beamforming is more realistic in a macro-cellular environment than in a NLOS environment, with receive diversity gain more prominent in a NLOS environment than in a macro-cellular environment.

108   SPACE-TIME PROCESSING FOR CDMA

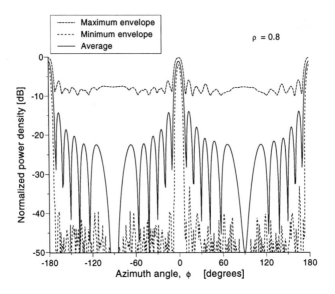

Figure 5.6.   ULA beam pattern with envelope correlation of $\rho_{ij} = 0.8$.

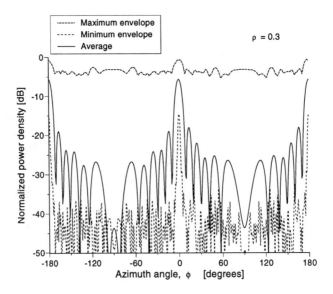

Figure 5.7.   ULA beam pattern with envelope correlation of $\rho_{ij} = 0.3$.

SMART ANTENNA TECHNIQUES 109

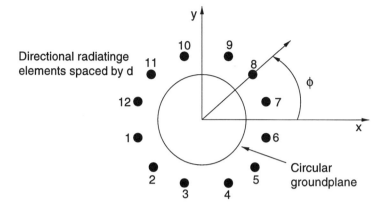

Figure 5.8. Circular array of $M_B = 12$ directional radiating elements.

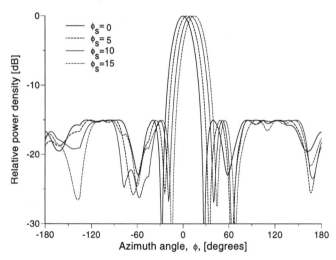

Figure 5.9. Radiation patterns for different scan angles for the $M_B = 12$ element circular array.

Element #	NLOS	LOS	Macro
4 and 9	0.825	0.95	0.99
5 and 8	0.625	0.875	0.96
6 and 7	0.525	0.825	0.94

Table 5.1. Correlation values for different channel models for a $M_B = 12$ element circular array.

110   SPACE-TIME PROCESSING FOR CDMA

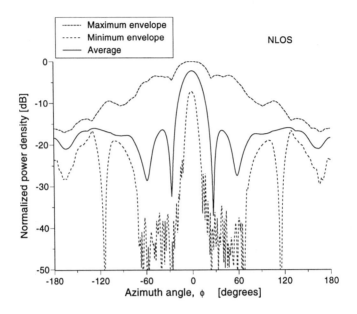

Figure 5.10.   Envelope for circular array with NLOS channel model correlation.

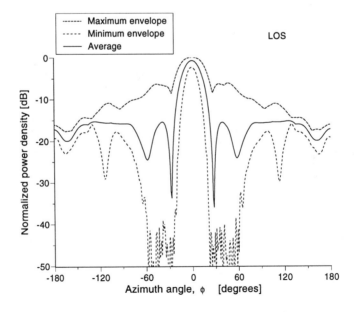

Figure 5.11.   Envelope for circular array with LOS channel model correlation.

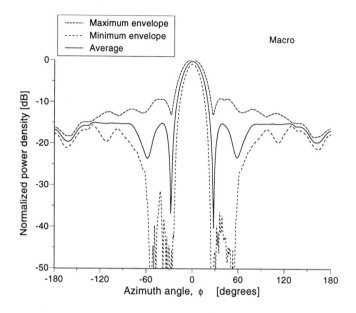

Figure 5.12. Envelope for circular array with macro-cell channel model correlation.

5.2 DIVERSITY

When the signal received at either the base station or the mobile experiences a deep fade, the quality of the signal is reduced and the certainty with which a bit decision can be made is reduced. However, if the receiver can be supplied with multiple copies of the same information, it should be conceivable that the receiver can choose the best replica and make a more reliable decision based on that signal, or it may even combine a number of the replicas in order to make a decision. The sources of these multiple signals can essentially be two-fold. Firstly, multiple replicas of the signal can be formed by the propagation path as described in Chapter 4. In this case, the receiver can employ a number of receiving antennas to receive multiple copies of the transmitted signal. This process is known as receive diversity. Alternatively, multiple replicas of the signal can be purposefully generated by the transmitter using multiple antennas. This is known as transmit diversity. In the following sections transmit and receive diversity techniques are explained in detail.

5.2.1 Transmit Diversity

For narrowband TDMA many techniques have been proposed to provide transmit diversity. These techniques can broadly be categorized as

- space-space transmit diversity (polarized antennas are often used to realize space-space transmit diversity),

- space-frequency transmit diversity and space-phase transmit diversity (the introduction of frequency offsets [137] and phase-sweeping [102, 107], to convert a frequency non-selective channel into a frequency selective channel is a technique used to realize such a transmit diversity scheme), and

- space-time transmit diversity (examples are FIR pulse shaping techniques imposing intentional inter-symbol interference (ISI) [323], delay diversity [242, 321], and space-time antenna-hopping (also known as round-robin antenna selection [188]) diversity schemes.)

Many of these techniques can easily be extended to CDMA. A general classification of transmit diversity techniques, for TDMA and CDMA, can be summarized as follows.

Transmit Diversity with Feedback [320, 213].
In these schemes, implicit or explicit (closed loop) information is fed from the receiver to the transmitter in order to configure the transmit diversity structure. Winters [320] considered switched diversity with feedback while Raleigh et al. [213] considered spatio-temporal-frequency water pouring, a technique based on channel feedback response. These techniques are limited in practice by vehicle movements and/or interference which causes a mismatch between the state of the channel perceived by the transmitter and that perceived by the receiver.

Transmit Diversity with Training Information [323, 242, 161].
Linear processing at the transmitter is used to distribute encoded and control (e.g. transmitter power) information to the antenna sub-sequences. Feedforward or training information is utilized to estimate the link from the transmitter to the receiver. The first scheme of this type was proposed by Wittneben [323], which includes the delay diversity scheme presented by Seshadri et al. [242] as a special case. Lu et al. [161] has considered a technique where channel reciprocity was assumed by assigning the same antenna weights to the transmitter and receiver via implicit (open loop) feedback.

Hybrid Feedback/Training Transmit Diversity.
In practice, the information update rate is slow, and channel reciprocity cannot be guaranteed because of vehicle movement and/or interference. Here, the feedback and training estimates are combined to compensate for the channel response at both the transmitter and receiver. In this way the best features of both open and closed loop channel state estimation are combined.

Blind Transmit Diversity [102, 107, 42, 242, 315].
No feedback or feed-forward information is required. Instead, blind transmit

diversity exploits the use of multiple transmit antennas combined with channel coding to achieve diversity. An example of this approach is to combine phase sweeping transmitter diversity [102] with channel coding [107]. Here a small frequency offset is introduced on one of the antennas to create fast fading. Another scheme is to encode the information by a channel code and then to transmit the code symbols using different antennas in an orthogonal manner. This can be achieved by either considering frequency multiplexing [42], or time multiplexing [242]. Also in the case of CDMA, orthogonal spreading sequences can be assigned to the different transmitting antennas [315]. When appropriate channel coding is employed, it is possible to relax the orthogonality requirement, with the benefit of achieving diversity and coding gain.

Two promising transmit diversity techniques well suited to CDMA are discussed in more detail below. These are code-division transmit diversity (CDTD) and time-division transmit diversity (TDTD).

5.2.1.1 Code-Division Transmit Diversity (CDTD).
In CDMA it is desirable to transmit orthogonal signals to different users in the downlink and to simultaneously maintain a fixed number of user channels. Clearly both these requirements cannot be met, since the number of available orthogonal channels is fixed. For this reason two main approaches are followed in transmit diversity for CDMA. These are orthogonal (O-CDTD) and non-orthogonal (NO-CDTD) CDTD. Both O-CDTD and NO-CDTD have the receive diversity property of soft-failure[1].

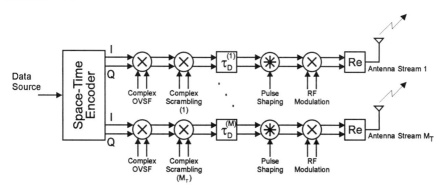

Figure 5.13. Block diagram of a general CDTD single user system.

Figure 5.13 illustrates how the WCDMA system described in Appendix G, and implemented by the accompanying simulation software, can be modified to include downlink CDTD. The main components are the space-time encoder, complex spreader, modulator and transmit antennas. Data modulation per

114 SPACE-TIME PROCESSING FOR CDMA

antenna is QPSK, where the space-time encoded downlink data is mapped to the I and Q branches. The data is spread by a combination of complex (orthogonal and non-orthogonal) variable spreading and complex scrambling codes. In both O-CDTD and NO-CDTD the symbol rate on each transmit antenna M_T is reduced by a factor $1/M_T$ to ensure that the data bits are evenly distributed to each transmit antenna element.

For O-CDTD, different complex orthogonal spreading codes are assigned to every antenna. This maintains the orthogonality between the two output streams, and hence self-interference is eliminated in flat fading. For NO-CDTD, the same complex spreading code is assigned to every antenna with an intentional delay between each antenna element. For this reason NO-CDTD is also known as delayed CDTD. Typical non-orthogonal spreading sequences used for NO-CDTD are Gold sequences. The advantage of Gold sequences are that they maximize the number of spreading sequences, but compromises orthogonality due to self-interference.

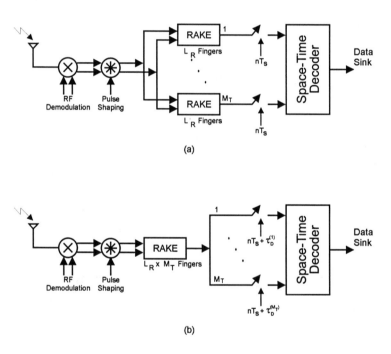

Figure 5.14. Single user CDTD receiver block diagram: a) O-CDTD b) NO-CDTD.

The general receiver structures for O-CDTD and NO-CDTD are shown in Figures 5.14(a) and 5.14(b), respectively. With reference to Figure 5.14(a), a total of M_T RAKE receivers, each with L_R fingers, are employed for O-CDTD. Each of the M_T RAKE receivers is trained on the spreading sequence associated with the corresponding transmit antenna. The M_T complex outputs are then sampled at the symbol rate, T_s, and passed to the space-time decoder.

In the NO-CDTD receiver, shown in Figure 5.14(b), a single RAKE receiver with $L_R \times M_T$ fingers is used. Since the same spreading sequence is used at the transmitter for all transmit antenna paths, *a priori* information of the time delays is needed. These delays are then used for RAKE post-processing. At the output of the RAKE combiner M_T complex samples (sampled at $t = nT_s + \tau_D^m$) are formed, and processed by the space-time decoder.

This increase in downlink MAI can be compensated for by additional RAKE fingers in the mobile receiver. In a typical mobile communication channel, where the scattering environment induces non-orthogonality on the spreading sequences, NO-CDTD should perform better since it is well known that non-orthogonal spreading sequences performance better than orthogonal spreading sequences in such an environment.

At the NO-CDTD receiver the most important extension to the receiver of Figure 2.10 is the addition of M_T RAKE fingers. This is especially important with space-time coding which attempts to exploit the degrees-of-freedom of the system more optimally. In general channel estimation is required to set up the RAKE receiver and to perform diversity reception. It is important to note that for NO-CDTD the pilot signal is also split and transmitted simultaneously on the available antennas which allows coherent detection of the signals received from all the transmit antennas. The data is processed using a RAKE finger with parallel processing capability. All the transmitted signal streams are received simultaneously at the same delay (for a given multi-path components), hence no additional buffering and skewing of data is necessary. This significantly reduces the hardware complexity/cost associated with NO-CDTD implementation.

5.2.1.2 Time-Division Transmit Diversity (TDTD).

Figure 5.15 illustrates the general TDTD structure for a single user in the downlink. Again an extension of the UMTS-like system described in Appendix G is depicted. TDTD for CDMA can be implemented as

Round-robin TDTD (RR-TDTD).

This scheme can be implemented by time-orthogonal (sharing) by using pseudo-random antenna hopping (round-robin) sequencing. For example, dual antenna time-switched RR-TDTD can be implemented by transmitting consecutive slots of the downlink by two separate antennas. After scrambling, the spread time slots can be switched consecutively to each antenna. The other TDTD users of the system may have different switching patterns in order to reduce the peak transmit power and peak to average power ratio in each power amplifier.

Antenna Selection TDTD (AS-TDTD).

The transmit antennas can be determined more optimal, by using feedback from every mobile to the base station to employ closed-loop antenna selection. In CDMA, in general, the control loop delay can be kept well within the channel coherence time to enable efficient use of power control and antenna selection loops.

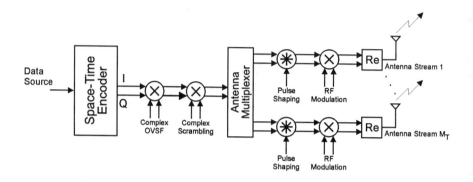

Figure 5.15. Block diagram of the space-time coded time-division transmit diversity (TDTD) system.

The main difference between RR-TDTD and AS-TDTD is that the distribution of the encoded bits in AS-TDTD is more selective. In other words, depending on the feedback information, the encoded bits are transmitted only from the best antenna. In [111], feedback signaling has been employed to simultaneously perform fast closed loop power control for downlink diversity.

The drawbacks associated with TDTD are that

- TDTD does not provide soft-failure,

- the channel decoder is not true space-time since it is unable to incorporate any space diversity information during the decoding process, and

- the channel has to be re-estimated on a slot-by-slot basis.

5.2.1.3 CDTD and TDTD with Pre-RAKE Combining. Figure 5.16 illustrates a general pre-RAKE combining TDTD transmitter. This transmitter is based on the CDMA pre-RAKE combining strategy by Jeong *et al.* [118]. The principle of operation is that the transmitted pre-RAKE signal is a time-reversed replica of the channel impulse response. In this way space and path diversity is possible at the mobile receiver without any conventional receive diversity techniques. The transmitter is based on two diversity principles, pre-RAKE (realized by the tapped delay line) and space diversity (realized by the multiple transmit antennas). A similar extension to CDTD is possible.

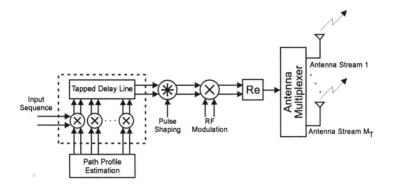

Figure 5.16. Block diagram of the pre-RAKE combining scheme for TDTD.

5.2.2 Receive Diversity

A diversity technique requires a number of signal transmission paths, called diversity branches, that carry the same information but have, ideally, uncorrelated multi-path fading, and a circuit to combine the received signals or to select one of them. This definition of diversity differs from beamforming where it is normally assumed that the signals arriving at the antenna array are perfectly correlated. Whenever the signals on the various antenna elements are not perfectly correlated, the beamforming pattern is influenced in a detrimental way (see Section 5.1.4). At some point, due to a lack of correlation, the beam pattern will revert back to an omni-directional pattern. Lack of correlation is normally induced by the environment through which the received signal is propagated and also due to the spacing of the antenna elements. Whenever the correlation, ρ_{ij}, between branch i and j, is less than perfect, that is $\rho_{ij} < 1$, there will be some diversity gain present in the system (our discussion will be limited to spatial diversity, but other types of diversity are also possible, see i.e. [142]). Diversity combining is different from antenna array processing and beamforming in that it combines signals at baseband or at IF to increase the signal level without affecting the individual antenna pattern. Beamforming techniques, on the other hand, exploit the differential phase between different antennas normally at RF level to modify the antenna pattern of the whole array. In this arrangement, once the signals are combined, the whole of the array has a single antenna pattern.

With diversity systems, we can therefore utilize the fact that the signals arriving at different locations fade at different rates. A system employing a diversity combiner uses signals induced on various antennas placed a few wavelengths apart at different locations and combines these signals in one of many ways [142]. For example, an equal gain combiner adjusts the phases of the desired signals and combines them in-phase after equal weighting. The MRC applies weights in proportion to the SNR and combines the weighted signals

118   SPACE-TIME PROCESSING FOR CDMA

in-phase, whereas a selection diversity combiner selects the signal from one of the antennas for processing. The selection may be based upon the power of the desired signal, the total power, or the SINR available at each antenna. The selection based upon the SINR is most effective in combating co-channel interference, whereas the equal gain combiner provides the lowest outage probability (the probability that the channel is not suitable for use). In theory, increasing the number of antennas results in greater reduction in channel fading. However, in practice, 10-20 antennas seem to provide satisfactory results [296].

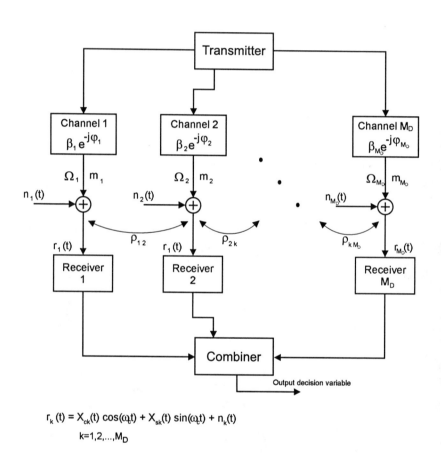

Figure 5.17.   Basic structure of a diversity system.

As indicated in Figure 5.17, let the received signal in the kth channel of the M_D-branch diversity system have power Ω_k, Nakagami fading parameter m_k, and correlation between the kth and $(k+1)$th branch be $\rho_{k(k+1)}$. It is assumed that each channel is frequency non-selective and slowly fading with channel attenuation factors $\{\beta_k\}, k = 1, \cdots, M_D$ having Nakagami-distributed envelope statistics, given by

$$p_{\beta_k}(\beta_k) = \frac{2}{\Gamma(m_k)} \left(\frac{m_k}{\Omega_k}\right)^{m_k} \beta_k^{2m_k-1} e^{-m_k \beta_k^2/\Omega_k}, \qquad (5.20)$$

with $\Omega_k = E\{\beta_k^2\}$. The fading *power*, $S_k = \beta_k^2$, therefore has a gamma distribution [2]. It is informative to note that Rayleigh fading is equivalent to Nakagami fading with fading parameters $m_k = 1$ and the one-sided exponential distribution when $m_k = 0.5$. It can also model the Rician distribution with sufficient accuracy by setting [255, 205]

$$m_k = \frac{1}{1 - \left(\frac{\mathcal{K}_k}{1+\mathcal{K}_k}\right)^2}, \qquad (5.21)$$

where $\mathcal{K}_k$ denotes the Rice factor (average direct power/average scattered power).

The signal in each channel is corrupted by an additive zero-mean white Gaussian noise process, assumed to be mutually statistically independent, with identical autocorrelation functions (i.e. powers). The received signal can then be expressed as

$$\begin{aligned} r_k(t) &= X_{ck}(t)\cos(\omega_c t) + X_{sk}(t)\sin(\omega_c t) + n_k(t), \\ k &= 1, 2, \cdots, M_D, \end{aligned} \qquad (5.22)$$

where $X_{ck}(t)$ and $X_{sk}(t)$ are zero mean stationary Gaussian random processes, represented by $N(0, \Omega_k/2)$, with $E\{X_{ck}^2\} = E\{X_{sk}^2\} = \Omega_k/2$ and $n_k(t)$ white Gaussian noise.

When binary signalling is assumed, the optimum demodulator for the signal received from the kth channel consists of two matched filters, one for each transmitted symbol. Following the matched filters is a combiner that forms the two decision variables. The combiner that achieves the best performance is one in which each matched filter output is multiplied by the corresponding complex-valued (conjugate) channel gain. The effect of this multiplication is to compensate for the channel phase shifts and to weight the signal by a factor that is proportional to the signal strength. Thus, a strong signal carries a larger weight than a weak signal. After the complex-valued weighting operation is performed, two sums are formed. One consists of the real parts of the weighted outputs from the matched filters corresponding to a transmitted 0. The second consists of the real part of the weighted outputs from the matched filters corresponding to a transmitted 1. This optimum combiner is called a maximal ratio combiner. Of course the realization of this optimum combiner is based on the assumption that the channel attenuations and phase shifts are perfectly known. Under these assumptions, the pdf of correlated multivariate gamma distributions (the maximum ratio combined signal powers on each diversity branch) is derived in Appendix D as a general result.

As an example of diversity performance consider Figure 5.18, which was obtained from

$$P_e = \left[\frac{1}{2}(1-\mu_k)\right]^{M_D} \sum_{k=0}^{M_D-1} \binom{M_D-1+k}{k} \left[\frac{1}{2}(1+\mu_k)\right]^k, \quad (5.23)$$

where

$$\mu_k = \sqrt{\frac{\overline{\gamma}_k}{1+\overline{\gamma}_k}}, \quad \gamma_b = M_D \cdot \overline{\gamma}_k \quad \text{and} \quad \overline{\gamma}_k = \Omega_k \cdot \frac{E_b}{N_0}.$$

Equation (5.23) is the standard expression for Rayleigh faded MRC [205] (the average error rate, P_e can also be verified in the WCDMA Simulation Environment described in Appendix H). Here the average SNR per branch is given by $\overline{\gamma}_k$ and the total average SNR per bit by γ_b. From this graph it is clear that, as the number of diversity branches increases, the system performance increases (i.e. average error rate decreases). What is also of significant importance to note, is that, as the number of diversity branches increases, the performance converges towards the unfaded, AWGN performance.

5.3 SECTORIZATION

The best known and simplest space-time processing technique used must surely be sectorization. Almost all cellular system use sectorization, primarily to reduce co-channel interference and to increase capacity. Typically, cells are divided into 3 sectors which are 120° wide or 6 sectors which are 60° wide [149]. Using this approach, it is possible to re-use spectral resources more frequently, thus allowing a larger traffic density to be supported in a cell. A typical implementation of a three beam sectorization scheme is shown in Figure 5.19(a), where three beams are deployed. From the figure it is important to note that the increase in system capacity achieved with sectorization comes at the price of increased system hardware in the form of radio channel units (RCU) required for each sector, as well as extra trunking resources. As CDMA systems are interference limited with matched filter detection, sectorization will linearly increase the capacity of the overall cellular system provided that the interference is uniformly distributed in angle. Thus, a three sector antenna system can increase the system capacity approximately three times.

5.4 SWITCHED-BEAM SMART ANTENNAS

An alternative strategy to using a beamforming array described in Section 5.1.3, is to use a switched-beam smart antenna system. Switched-beam smart antenna

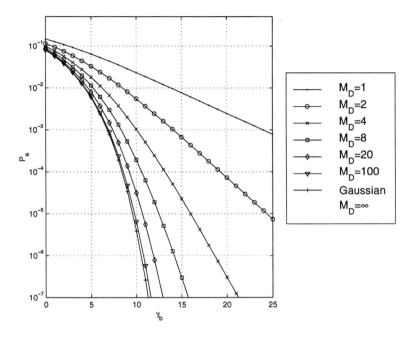

Figure 5.18. Performance of MRC diversity system.

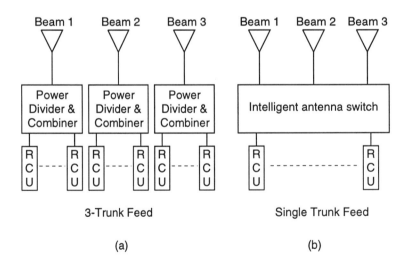

Figure 5.19. Sectorization and switched-beam smart antenna systems.

systems consist of multiple narrow beam directional antennas controlled by a beam-selection algorithm [108]. A typical switched-beam antenna system is shown in Figure 5.19(b) alongside a basic sectorization system. In this figure, the intelligence of the smart antenna system resides in the antenna switch which incorporates signal processing capability to locate a subscriber within a beam and control the switching such that the beam receiving the subscriber with the best SINR is connected to the RCU [155].

The conceptual operation of a switched-beam smart antenna is relatively simple. Narrow beams are created using physically directive antenna elements to create aperture and gain [27]. If the received signal SINR falls below a predefined value during a connection, the antenna switch will select a different antenna beam to connect the subscriber to the RCU. The directive nature of narrow beam antennas ensure that, in a given system, the level of co-channel interference experienced from and by neighboring cells is much less, on average, than that experienced when using conventional wide-coverage antennas [260]. Furthermore, the intelligent antenna switch increases the trunking efficiency of the system leading to an overall increase in system capacity [155].

Concentrating on interference reduction, if it is assumed that interfering signals are uniformly distributed in angle, the minimum SINR of an antenna system with M_{sb} beams and a frequency re-use factor, r_c, is [155]

$$SINR = \frac{3}{2} r_c^2 M_{sb}. \tag{5.24}$$

As an example of the interference reduction effects of the switched-beam system, it is interesting to note that a 12-beam, $r_c = 4$ switched-beam system has a SINR of 24.5 dB which is greater than the SINR of a three-beam, $r_c = 7$ conventional sectorization system. Thus, the switched-beam system provides a capacity increase of 7/4.

In addition to the increased capacity that may be obtained due to interference reduction, switched-beam systems also increase trunking efficiency. It is well known that the highest trunking efficiency is obtained when all channels assigned to a cell are available to every subscriber in the cell (from Figure 5.19 it is clear that this is not the case for sectorization). The measure of trunk capacity is Erlang, where one Erlang is an hour of channel traffic at a specific blocking rate, B_E, per hour. Specifically, if a 1% blocking rate and a subscriber usage of 0.03 Erlang is assumed [143], the number of subscribers per channel for a switched-beam system can be approximated by

$$N_{S/CH} \approx 28.5 \tanh(0.07 N_{ch}) - 0.047 N_{ch}^2 \exp(-0.07 N_{ch}), \tag{5.25}$$

where N_{ch} denotes the number of channels for a trunk. From this equation it follows that converting, for example, a three sector, three trunk cell to a three sector, one trunk cell with intelligent antenna switching would eliminate sector-to-sector hand over as well as increase capacity by 26% [155].

In order to determine the overall capacity of a switched-beam system, the number of channels per cell is denoted by [155]

$$M_{ch} = \frac{N_T}{r_c}, \qquad (5.26)$$

where N_T is the total number of channels in a frequency re-use cluster. Substituting (5.24) in (5.26) results in

$$M_{ch} = \frac{N_T}{\sqrt{\frac{2SINR}{3M_{sb}}}}, \qquad (5.27)$$

with the number of subscriber per cell equal to

$$K_{sub} = N_{S/CH} M_{ch}, \qquad (5.28)$$

and the number of channels per trunk equal to

$$N_{CH} = M_{ch}/n_{ch}, \qquad (5.29)$$

where n_{ch} denotes the number of trunks per cell.

5.5 SUMMARY

This chapter classified, defined and discussed the various aspects of smart antenna systems. The main smart antenna techniques, beamforming, transmit- and receive diversity, sectorization and switched beamforimg, were discussed in detail. Beamforming algorithms have been discussed with emphasis on algorithms applicable to CDMA. In the following chapters, the background developed up to now will be used for performance analysis (Chapter 6), multiuser detection (Chapter 7) and space-time coding (Chapter 8).

Notes

1. Soft-failure states that, should one of the receive chains fail, and the other chain is operational, the performance loss is of the order of the diversity gain. In other words, the signal may still be detected, but with inferior quality.

6 SMART ANTENNA PERFORMANCE

One of the main performance criteria in a digital communications system is the average bit error rate of the system as a function of the received SNR. As an introduction into the analytical evaluation of space-time processing systems, this chapter uses the system model presented in Chapter 2 in conjunction with the space-time channel model presented in Chapter 4 to evaluate the performance of space-time CDMA systems excluding multiuser detection and space-time coding. The latter two system's performance is presented in Chapters 7 and 8 respectively. Specifically, the performance of a CDMA system in a beamforming environment (as described in Section 6.1), a diversity environment (as described in Section 6.2) and then a combination of beamforming and diversity is considered. Furthermore, analytical results showing the influence of the space-time channel models discussed in Chapter 4 are presented are the basic trade-offs between diversity and beamforming are discussed.

6.1 BEAMFORMING ARRAY PERFORMANCE

In the analysis presented in this section, a single cell, star connected network with a base station situated in the middle of the cell is assumed. The mobile users are distributed according to (3.13) throughout the cell. Each of the mobile users is surrounded by local scattering elements with a Gaussian density causing multi-path signals to be generated with DOA as in (4.14). In the uplink

K users transmit signals over time-invariant multi-path channels, each with L multi-path elements, to the base station.

With reference to the the general space-time processing system model presented in Figures 2.9 and 2.10 only one transmit antenna ($M_T = 1$), one receive diversity branch ($M_D = 1$) and M_B beamforming antenna elements are considered with an L_R tap RAKE receiver. The M_B element beamformer is assumed to form an antenna pattern that is matched to the pdf of the DOA of the reference user's signal, which can be accomplished with one of the spatial filtering techniques described in Section 5.1.3. This beamforming approach yields an antenna system that maximizes the received SNR of the reference user's signal. Also, the beamformer used in this analysis does not implement null-steering to minimize interference from specific high power interferers. Such algorithms to determine the set of weights, $\mathbf{w}^{(k)}$, are treated in Section 5.1.3. It should be noted that the analytic derivation of the BER performance is not dependent on specific antenna weight values. Should a specific antenna synthesis method be used to determine a set of weights, these weights can be substituted in the performance analysis to yield performance results for specific antenna patterns. Furthermore, the antenna array elements are assumed to be sufficiently closely spaced to ensure high correlation at each antenna element. If the received signals are not correlated, typical effects as described in Chapter 5 can be expected in the antenna beam pattern.

Following the model derived in Section 2.4, our current analysis assumes that $P = 0$, that is, only one symbol is being detected (unlike the multiuser case where the detection process is dependent on $P + 1$ detected bits), and therefore, (2.9) reduces to

$$\hat{c}_l^{(k)}(t) = \beta_l^{(k)} \exp(j\varphi_l^{(k)}) u(t/T_s). \quad (6.1)$$

In all cases considered here, the number of RAKE fingers, L_R, can be equal, greater or smaller than the number of received multi-path signals L. The received signal is therefore first processed by the beamformer after which the reference signal is despread. From (2.16), the decision variable of the i-th bit of user j at the RAKE receiver output can be written as [62]

$$\zeta^{(j)} = \sum_{n=1}^{L_R} \left\{ S_n^{(j)}(i) + I_{si_n}^{(j)}(i) + I_{main}^{(j)}(i) + \eta_n^{(j)}(i) \right\}. \quad (6.2)$$

To arrive at an expression for the BER of a beamforming system with a RAKE receiver, we need to calculate the signal power, U_S^2, and the total interference power, σ_T^2. With these variables known, the received SNR is

$$\text{SNR} = \frac{U_S^2}{\sigma_T^2}. \quad (6.3)$$

From (2.22), the ULA beamformer for the reference user ($k = j$) will calculate

$$\|\mathbf{w}^{(j)}\|^2 = \left\| \left(1, \exp(j\chi d_x \cos\phi^{(j)}), \cdots, \exp(j\chi(M_B - 1)d_x \cos\phi^{(j)})\right)^T \right\|^2$$
$$= \left(\mathbf{w}^{(j)}\right)^H \mathbf{w}_j^{(k)} = \sum_{i=1}^{M_B} w_i^* w_i = M_B. \quad (6.4)$$

To simplify the analysis, it is assumed that the sum of all interference terms of (6.2) are Gaussian distributed [59, 62, 206, 285]. This assumption has been shown to be accurate, even for small values of K when the BER is 10^{-3} or greater. Therefore, expanding on the results of [62, 59] to include beamforming, the variance of each interference term (equations (2.18)-(2.20)) on the nth RAKE tap, conditioned on the fading parameter, $\beta_n^{(j)}$, can be written as

$$\left(\sigma_{si_n}^{(j)}\right)^2 = \frac{E_b T_s}{4N} \sum_{\substack{l=1 \\ l \neq n}}^{L} \left(\beta_n^{(j)} \|\mathbf{w}^{(j)}\| E\left\{\overline{\mathcal{R}}^{(jj)}\right\}\right)^2 \Omega_l^{(j)}, \quad (6.5)$$

$$\left(\sigma_{main}^{(j)}\right)^2 = \frac{E_b T_s}{6N} \sum_{\substack{k=1 \\ k \neq j}}^{K} \sum_{l=1}^{L} \left(\beta_n^{(j)} \|\mathbf{w}^{(k)}\| E\left\{\overline{\mathcal{R}}^{(jk)}\right\}\right)^2 \Omega_l^{(k)}, \quad (6.6)$$

$$\sigma^2 = \frac{T_s N_0}{4} \cdot \left(\beta_n^{(j)}\right)^2, \quad (6.7)$$

where $\Omega_l^{(k)}$ denotes the average signal power of path l received from user k. This in turn yields a total interference term of

$$\sigma_T^2 = \sum_{n=1}^{L_R} \left(\left(\sigma_{si_n}^{(j)}\right)^2 + \left(\sigma_{mai_{qn}}^{(j)}\right)^2 + \sigma^2\right), \quad (6.8)$$

where L_R denotes the number of branches in the RAKE receiver. Furthermore, the desired signal output of the RAKE combining receiver can be written as

$$U_S = \sqrt{\frac{E_b T_s}{2}} \sum_{n=1}^{L_R} \|\mathbf{w}^{(j)}\| (\beta_n^{(j)})^2. \quad (6.9)$$

In (6.9), it is assumed that the RAKE receiver will recover the strongest (i.e. largest average received signal power) L_R multi-path components. For convenience sake, and without loss of generality, the strongest multi-path components are assumed to be the components that arrive at the receiver first (compare with the MIP in Appendix C).

The variance of the fading parameters of each interfering user, $E\left\{(\beta_l^{(k)})^2\right\}$ is equal to the average signal power received from that user, $\Omega_l^{(k)}$. This variance is not a function of the antenna array or steering vector, as the fading process is caused by physical scattering processes that occur at the mobile. The effect of the array is encapsulated in the spatial correlation parameter (given by (2.21)), which will be a minimum if the array has a null in the direction of a specific interfering multi-path signal arriving at the base station, thereby minimizing the average received signal power from the interfering multi-path component (see also Figure 5.5). It is assumed that the multi-path is characterized by an exponential MIP [62], i.e.

$$\Omega_l^{(k)} = \Omega_1^{(k)} e^{-(l-1)\delta_d}, \qquad \delta_d > 0 \qquad (6.10)$$

where $\Omega_1^{(k)}$ is the average signal strength corresponding to the first incoming path of user k and δ_d is the rate of average power decay.

For coherent demodulation, the BER conditioned on the instantaneous SNR, S, can be expressed as [325]

$$P_{e|S} = Q\left(\sqrt{\sigma_0 \cdot s}\right), \qquad (6.11)$$

where $Q(x) = \frac{1}{\sqrt{2\pi}} \int_x^\infty e^{-t^2/2} dt$ is the Q-function [2]. The output SNR (as defined in (6.3)), can be written in the form required by (6.11) as

$$S = \sum_{n=1}^{L_R} \left(\beta_n^{(j)}\right)^2, \qquad (6.12)$$

and

$$\sigma_0 = \left(\frac{1}{2N} \sum_{\substack{l=1 \\ l \neq n}}^{L} \left(E\left\{\overline{\mathcal{R}}^{(jj)}\right\}\right)^2 \Omega_l^{(j)} + \frac{1}{3N} \sum_{\substack{k=1 \\ k \neq j}}^{K} \sum_{l=1}^{L} \left(E\left\{\overline{\mathcal{R}}^{(jk)}\right\}\right)^2 \Omega_l^{(k)} \right.$$
$$\left. + \frac{N_0}{2E_b \|\mathbf{w}^{(j)}\|^2} \right)^{-1}, \qquad (6.13)$$

assuming that $\|\mathbf{w}^{(k)}\|^2$ is equal for all k.

To obtain the average BER, (6.11) must be averaged over the pdf of S. As discussed in Chapter 4, the distribution of S should accommodate different values of the fading parameter m for the different received paths. If we assume that the fading amplitude, $\beta_n^{(j)}$, is Nakagami distributed (as described in Section 5.2.2), the *power*, $\left(\beta_n^{(j)}\right)^2$, of the received fading amplitude will be gamma distributed. From (6.12) it is clear that we need the pdf of the sum of L_R gamma distributed random variables to obtain the average error rate. In Appendix D a general pdf for the sum of an arbitrary number of correlated gamma distributed random variables are derived, and repeated here in terms of the characteristic function

$$p_S(s) = \frac{1}{2\pi} \int_{-\infty}^{\infty} \Phi_S(t) e^{-its} dt, \quad (6.14)$$

where $\Phi_S(t)$ is the characteristic function defined in Appendix D. The BER for a beamforming system with a RAKE receiver can now be written as

$$P_e = \int_{-\infty}^{\infty} P_{e|S} p_S(s) ds, \quad (6.15)$$

which can be solved using numerical methods.

For the special case where the Nakagami fading parameter of all multi-path components are Rayleigh distributed and equal, i.e. $\{m_l\} = 1$, and the average signal strength $\{\Omega_l\} = \Omega$, (6.14) reduces to

$$p_S(s) = \frac{1}{(L_R - 1)! \Omega^{L_R}} s^{L_R - 1} \exp(-s/\Omega). \quad (6.16)$$

Furthermore, it can easily be shown that (6.11) reduces to

$$P_e = Q\left(\sqrt{\left(\frac{N_0}{2E_b} + \frac{(K-1)}{3N}\right)^{-1}}\right), \quad (6.17)$$

for perfect power control, no multi-path propagation and an omni-directional antenna. This is consistent with the result obtained by Pursley [206] under similar conditions.

6.1.1 Optimum Antenna Element Spacing

In this section a parameter for optimum ULA antenna element spacing, d_x, in a CDMA environment is derived by minimizing the total interference $E\{\sigma_T^2\}$, defined by (6.8). For a ULA, d_x is the only parameter that can be optimized to reduce the total interference (see Section 2.2), which is not necessarily in general true for any other antenna geometry. It is clear that if $E\{\sigma_T^2\}$ is a minimum, the system will have optimal performance. This will be the case when no MAI is present. In a beamforming environment, this situation will arise when the interferers, relative to the reference user, are in the antenna nulls, and consequently the spatial correlation $\mathrm{Re}\left[\left(\mathbf{w}^{(j)}\right)^H \mathbf{w}^{(k)}\right] = 0$. Our aim is therefore to optimize (minimize) $E\{\sigma_T^2\}$ in the presence of MAI.

It is well known that for a broadside ULA (constant steering vector), there is always a maximum gain at 90° [16]. Assume then that the desired user is at a spatially optimal position, $\phi_0^{(j)} = 90°$. This is arbitrary since it is assumed that the main beam of an adaptive antenna will always be on the desired user when one of the optimally spatial filtering algorithm (as described in Section 5.1.3) are used. In the derivation it is also assumed that the interfering users are distributed around the base station according to the user distribution pdf defined by (3.13), and repeated here for convenience

$$p_{\Phi_0^{(j)}}(\phi_0^{(j)}) = \frac{1}{A_{norm}} \left[1 + \sum_{l=1}^{N_{peak}} \varrho_l \left[\mathrm{rect}\left(\frac{\varpi_l \phi_0^{(j)}}{\pi} - \varsigma_l\right) \right. \right. \tag{6.18}$$

$$\left. \left. + \mathrm{rect}\left(\frac{\varpi_l \phi_0^{(j)}}{\pi} - \varsigma_l - 2\pi\right) \right] \cos^2(\varpi_l \phi_0^{(j)} - \varsigma_l) \right]$$

$$\forall\, 0 \leq \phi_0^{(j)} \leq 2\pi.$$

We can therefore derive the optimum ULA antenna element spacing parameter as

$$\rho_s = E\left\{\left(\mathrm{Re}\left[\left(\mathbf{w}^{(j)}\right)^H \mathbf{w}^{(k)}\right]\right)^2\right\}$$

$$= \int_0^{2\pi} \left(\mathrm{Re}\left[\sum_{n=0}^{M-1} \exp\left(j\chi n d_x \cos \phi_0^{(j)}\right)\right]\right)^2 p_{\Phi_0^{(j)}}\left(\phi_0^{(j)}\right) d\phi_0^{(j)}. \tag{6.19}$$

where (6.18) can be solved numerically. If a uniformly distributed DOA at the base station is assumed, that is $p_{\Phi_0^{(j)}}\left(\phi_0^{(j)}\right) = 1/2\pi$, (6.18) has the analytic solution

$$\rho_s = \frac{1}{2} \sum_{x_1=0}^{M_B-1} \sum_{x_2=0}^{M_B-1} [J_0(\epsilon(x_1+x_2)) + J_0(\epsilon(x_1-x_2))], \tag{6.20}$$

where J_0 is the zeroth order Bessel function [2] and $\epsilon = 2\pi/d_b$ with $d_x = \lambda/d_b$.

Figures 6.1 to 6.3 show the spatial correlation, ρ_s, as a function of d_b for different values of M_B and various user distributions. From the figures it is clear that, for a given number of antenna elements, the optimum value of $d_b = 2$ when uniform and clustered (at 180°) user distributions are considered. When

SMART ANTENNA PERFORMANCE   131

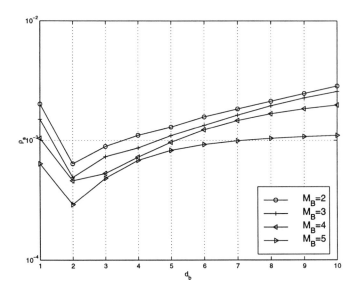

Figure 6.1. Influence of antenna spacing on total interference received by the base station for uniform user distribution (physical spacing, $d_x = \lambda/d_b$).

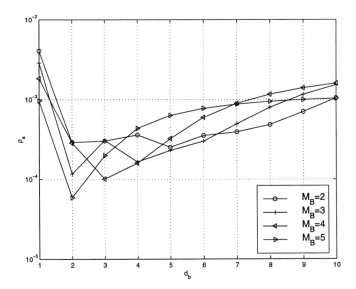

Figure 6.2. Influence of antenna spacing on total interference received by the base station for users clustered at 180° (physical spacing, $d_x = \lambda/d_b$).

132   SPACE-TIME PROCESSING FOR CDMA

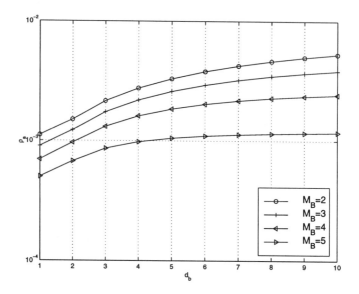

Figure 6.3. Influence of antenna spacing on total interference received by the base station for users clustered at 90° (physical spacing, $d_x = \lambda/d_b$).

the users are clustered at 90°, the optimum spacing is λ. However in this case, the correlation between the received fading envelopes is only 0.6 (using (4.17)) and, as is shown in Chapter 5, this correlation value would be too low for accurate DOA estimation and beamforming. Therefore, an optimum antenna element spacing for a ULA is $d_x = \frac{\lambda}{2}$ (which will also be used in the performance calculations of this chapter). The choice of an antenna spacing of $d_x = \frac{\lambda}{2}$ can further be substantiated by examining Figure 6.4. When the antenna elements are far apart ($d_x = 2\lambda$ and $d_x = \lambda$), a large number of grating lobes are present. In both cases, some of these very narrow but unsteerable lobes are pointed at the reference user, enabling very good performance when the users are clustered at 90° (see also Figure 6.3). However, when the users are more uniformly distributed, or even clustered over a wider region, the system performance will be degraded. When $d_x = \lambda/2$, the grating lobes disappear and a fairly narrow beam can be formed on the reference user leading to good performance irrespective of user distribution. As the antenna elements are moved closer together, the directivity of the array is reduced (it starts to approximate an omni-directional antenna) and consequently the interference rejection properties of the array is reduced (see for instance the case when $d_x = \lambda/8$ in Figure 6.4).

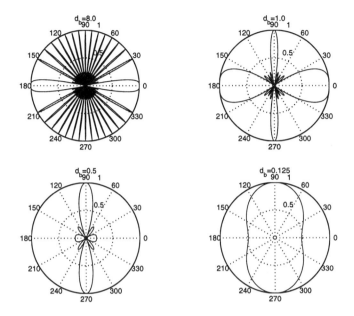

Figure 6.4. Influence of antenna spacing on the radiation pattern of a ULA (physical spacing as a fraction of λ given).

6.1.2 Numerical Results

Using (6.15) and the system parameters outlined in Table 6.1, the BER performance of a CDMA system using beamforming can be determined numerically under various physical and implementation conditions. In Figure 6.5, the influence of the beamforming antenna array size on the BER of a cellular CDMA system is shown. As would be expected, the BER performance of the system is better when the size of the beamformer is increased. This is due to the fact that larger beamforming arrays can synthesize narrower beams and thereby reduce the MAI seen by the reference user. As the BER probability is reduced by increasing the number of elements in the beamformer, the capacity of the cellular system is also increased. Figure 6.6 shows the increase in capacity that may be obtained in both LOS and NLOS environments (as described in Chapter 4) when beamforming is used. The system load is defined as $V = K/N$.

In addition to the influence of the size of the beamformer on the performance of the system, the effect of the spatial/temporal channel model used in the analysis can also be significant. Figures 6.7 and 6.8 show respectively the BER performance and capacity of the CDMA system as a function of the variation in temporal fading experienced by the reference user. This temporal fading is based on the model described in Chapter 4, and the figures clearly show that it is of the utmost importance to model accurately the spatial and temporal features of the channel.

Parameter	Simulation value
Spreading sequence length	$N = 64$
Operating environment	see Table 4.1
User distribution	uniform
Multi-path intensity profile	$\delta_d = 0$
Number of multi-path signals	$L = 5$
Number of users	$K = 5$
Number of RAKE fingers	$L_R = 2$
Number of beamforming elements	$M_B = 3$
Reference user location	$\phi_0 = 90°$
Temporal fading	$m = 5.76$ (LOS micro-cell)
	$m = 1$ (NLOS micro-cell)

Table 6.1. System parameters for numerical evaluation of BER performance.

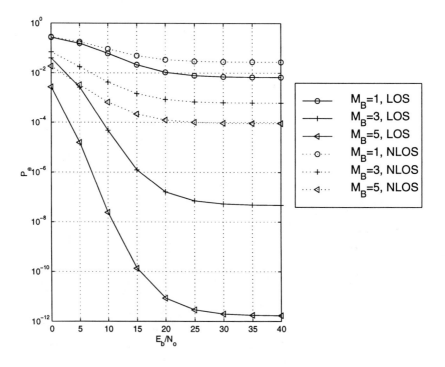

Figure 6.5. Influence of antenna array size on the BER of a cellular CDMA system including beamforming.

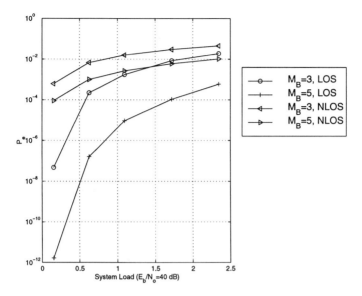

Figure 6.6. Influence of antenna array size on the capacity of a cellular CDMA system including beamforming.

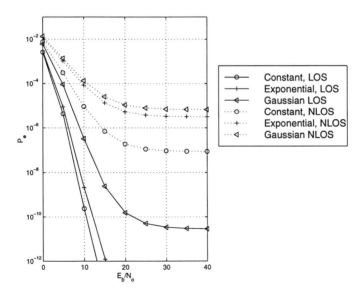

Figure 6.7. Influence of temporal fading on the BER of a cellular CDMA system including beamforming ($M_B = 3$).

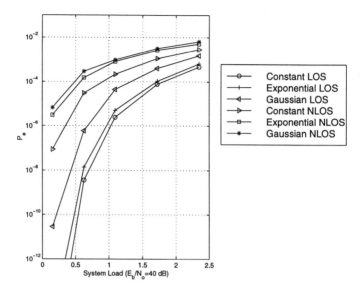

Figure 6.8. Influence of temporal fading on the capacity of a cellular CDMA system including beamforming ($M_B = 3$).

Based on the analytical results presented here, as well as the results in [157], the following general conclusions can be drawn regarding the use of a ULA beamformer in a cellular CDMA environment:

- The optimum overall system performance may be obtained using a combination of a RAKE receiver and a beamforming array.

- Increasing the number of fingers in the RAKE receiver leads to the largest increase in system performance. This increase in system performance is, however, dependent on the MIP.

- Increasing the number of elements in the antenna array leads to large increases in system performance, provided the users are approximately uniformly distributed in the cell.

- Antenna arrays are not suitable for areas where users are clustered closer in angle smaller than the minimum beam width of the array.

- When large arrays (or sectorized antennas with narrow sectors) are used, variations in the temporal fading distribution on each of the diversity branches significantly influence system performance. Accurate information on these distributions is required to predict overall system performance.

6.2 RECEIVE DIVERSITY PERFORMANCE

In this section, the BER performance of a MRC receive diversity system (described in detail in Chapter 5) with arbitrary correlated fading on each of the receive diversity branches will be determined. In beamforming applications it is assumed that the beamforming antenna array receives L uncorrelated multi-path components, with the received signal envelope correlation at each of the array elements equal to one. The same signal is therefore received by all elements of the beamformer (see also Chapter 5). In Figure 6.9, the difference between beamforming and receive diversity is graphically depicted. In beamforming, it is assumed that an identical set of L uncorrelated multi-path components arrive at all elements in the beamforming array. The correlation between the received signal envelopes at each element in the array is therefore equal to one as each composite multi-path signal consist of exactly the same L uncorrelated signals.

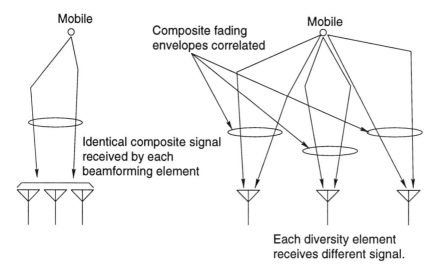

Figure 6.9. Composition of beamforming and diversity signals.

In the case of diversity, the situation changes since the elements of the diversity array are separated by a larger physical distance. This means that the L uncorrelated multi-path signals arriving at the first element in the diversity array is no longer the exact same set of uncorrelated multi-path components received by the other elements in the array. In fact, when some of the elements of the diversity array are separated by a large distance (typically 20λ or more), the L uncorrelated multi-path signals received at one element in the diversity array are completely different to the L multi-path signals received at any other element in the diversity array. A physical interpretation of this is that the scattering areas generating the multi-path echoes for one element in the diversity array are completely different from the scattering areas generating the multi-path components received at any other element in the diversity array. Furthermore, as two or more diversity elements start to receive L_s, where $L_s < L$, identical multi-path components, the correlation between the composite fading envelopes received increases from zero to some value less than one. Ultimately, if $L_s = L$, the correlation between the composite fading envelopes received at each element in the diversity system will equal one.

In order to determine the BER of a MRC diversity system, it is necessary to determine the pdf of the SNR at the output of the MRC receiver. The SNR pdf is a function of

- the characteristics of the composite signal received at each diversity branch (the effective value of the Nakagami fading parameter m). It has been shown [325, 59] that coherently combined, uncorrelated multi-path components result in a composite signal envelope with effective Nakagami fading parameter, m_{eff}, given by

$$m_{eff} = \sum_{l=1}^{L_R} m_l. \qquad (6.21)$$

If an antenna element in a diversity array receives three multi-path echoes, each with a Nakagami fading parameter $m_l = 1$ and coherently combines these signals using a RAKE combiner, the composite fading signal will have an effective Nakagami parameter $m_{eff} = 3$. Assuming that all other diversity branches receive and optimally combine the same number of uncorrelated multi-path components, each of the elements of the diversity array would effectively be receiving a composite multi-path signal which will have a Nakagami fading parameter of $m_{eff} = 3$, and

- the relation (correlation) between the signals received at each branch. The correlation between the composite fading envelopes received at each diversity branch is a function of the antenna height and also of the scattering environment as described in Chapters 3 and 4.

Moving on from the conceptual discussion above, the decision variable of a diversity system (see (2.16)) can be written as

$$\zeta^{(j)} = \sum_{m_D=1}^{M_D} \zeta_{m_D}^{(j)}, \qquad (6.22)$$

where each

$$\zeta_{m_D}^{(j)} = S_{m_D}^{(j)} + I_{si_{m_D}}^{(j)} + I_{mai_{m_D}}^{(j)} + n_{m_D}, \qquad (6.23)$$

and represents the output of the RAKE receiver on a specific diversity branch. This output of a diversity branch has the same form as the output of the beamformer described in (6.2), with $L_R = 1$, $M_B = 1$ and $m = m_{eff}$. From (6.22), it is clear that the outputs of the M_D RAKE receivers are again coherently combined as was done by the RAKE combiner used with the beamformer. Thus, the analytic derivation of the BER as presented in Section 6.1 is still valid. Therefore, from [325] and (6.11) it follows readily that the conditional BER of a diversity system is also

$$P_{e|S} = Q\left(\sqrt{\sigma_0 \cdot s}\right), \qquad (6.24)$$

where the received signal power random variable, S, will have a different pdf in (6.24) as opposed to (6.11). Specifically for receive diversity with MRC, the pdf of S is given in Appendix D, with the transformation of $M = M_D$ and $m = m_{eff}$. For Rayleigh fading (Nakagami fading parameter $m = 1$) with equal path strength ($\Omega_l = \Omega$) and equal correlation, a special case for the pdf of S follows as

$$p_S(s) = \frac{1}{\Omega^2 \Gamma(M_D)} \left(\frac{s}{\Omega^2}\right)^{M_D-1} \qquad (6.25)$$

$$\times \frac{\exp\left(-\frac{s}{(1-\rho)\Omega^2}\right) \cdot {}_1F_1\left(1, M_D, \frac{\rho M_D s}{(1-\rho)(1-\rho+\rho M_D)\Omega^2}\right)}{(1-\rho)^{(M_D-1)}(1-\rho+\rho M_D)},$$

where M_D the number of MRC diversity branches.

With reference to (6.24), the unknown variables required to determine the BER performance of a diversity system are the interference term σ_0 and the correlation matrix. The interference term has been defined in (6.12) for CDMA with beamforming. However, as we have shown, the analysis is equally valid for receive diversity with $M_B = 1$ elements. Using the pdf described by (6.14) in conjunction with (6.24), the BER performance can easily be determined using the methods described in Section 6.1 and numerical integration.

6.2.1 Numerical Results

Using this method, the BER performance of a receive diversity system under LOS and NLOS conditions with respectively $M_D = 3$ and $M_D = 5$ receive diversity branches (no additional RAKE fingers) are shown in Figures 6.10 and 6.11. In both cases it is clear that the correlation between the received signals on the diversity branches has a significant impact on the BER performance of the system. Using a constant correlation model and correlation values derived from Figure 4.8, the figures indicate that variations in the correlation between received signal envelopes as a function of the user's position or other spatial parameters must be taken into account when determining the capacity of a diversity system. This is again evident from Figures 6.11 and 6.12 where the influence of the number of diversity branches as well as the correlation between the signals received at the various branches is clearly shown.

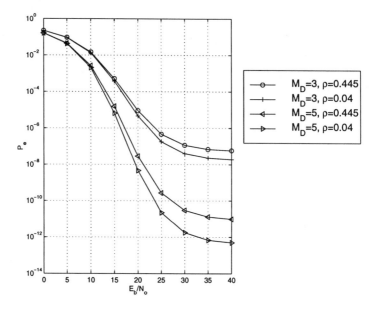

Figure 6.10. BER of diversity system under LOS conditions.

6.3 COMBINED DIVERSITY AND BEAMFORMING PERFORMANCE

As has been stated in Section 6.2, the BER analysis of an MRC diversity system is very similar to the analysis of a beamformer. Specifically, the only difference lies in the interference term σ_0. Thus, in order to determine the BER performance of a combined beamforming and diversity system, equations (6.24) and (D.12) can still be used. In this case, each of the M_D diversity branches consists of an M_B-element beamformer. Thus, when the interference term seen by each element in a diversity system is determined, (6.12) is used. The resulting interference term is then used in (6.24) for each of the M_D diversity branches. Furthermore, for the combined diversity and beamforming system, the pdf of the output SNR is a gamma distribution with M_D orders of freedom and where the Nakagami parameter at each diversity branch is determined by (6.21).

6.3.1 Numerical Results

Determining the BER performance numerically, it is clear that a combination of diversity and beamforming provides the best performance results, albeit with the highest implementation cost. Figures 6.13 and 6.14 depict the BER performance of a combined diversity and beamforming system. Specifically,

SMART ANTENNA PERFORMANCE   141

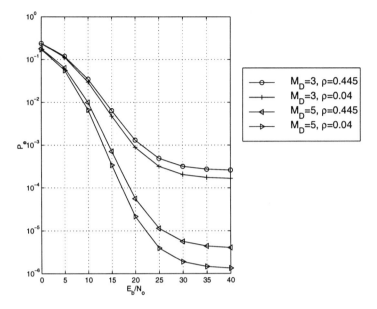

Figure 6.11.   BER of diversity system under NLOS conditions.

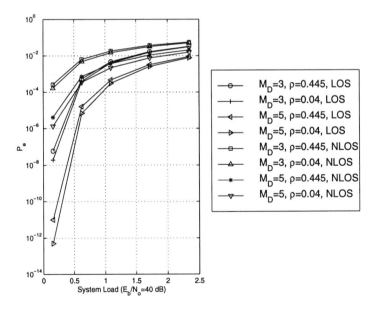

Figure 6.12.   Capacity of diversity system under LOS and NLOS conditions.

142   SPACE-TIME PROCESSING FOR CDMA

each beamformer is assumed to consist of three antenna elements ($M_B = 3$) and the complete beamformer forms one branch of the diversity system. Thus, when $M_D = 3$ diversity branches are present, a total of 3×3 antennas is used. Again, the effect of correlation on the system capacity can be seen in Figure 6.15.

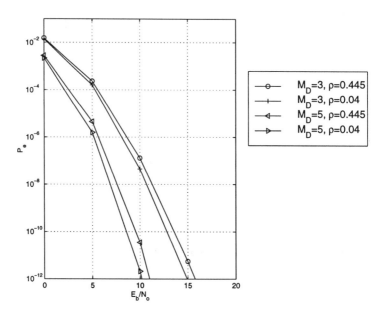

Figure 6.13.   BER of diversity system under LOS conditions.

6.4   CHOOSING A SPATIAL PROCESSING TECHNIQUE

Based on the derivation of the BER performance for beamforming, diversity and combined diversity and beamforming systems presented in the preceding sections, the natural question that arises is under what circumstances each of the techniques would provide the most system gain. In this section, the two cases of LOS and NLOS propagation are first treated separately, after which some general conclusions are drawn. As a basis for this section, the simulation parameters described in Table 6.2.

A comparison between the BER performance of a beamforming and a diversity system is shown in Figure 6.16. Both system use the same number of antennas ($M_B = M_D = 3$ in this case), meaning that the implementation cost of both systems is approximately equal. Considering the performance of the diversity system first, it is clear that the correlation between the diversity elements influences the achievable BER performance. Thus, when designing the diversity system, the assumption that the signals received at each element are uncorrelated cannot be used in BER computations. Specifically, the presence of

SMART ANTENNA PERFORMANCE 143

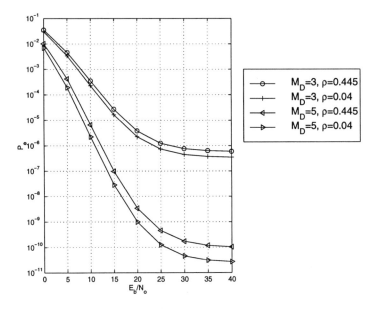

Figure 6.14. BER of diversity system under NLOS conditions.

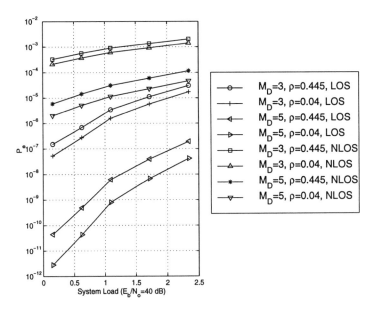

Figure 6.15. Capacity of diversity system under LOS and NLOS conditions.

Parameter	Simulation value
Environment	LOS and NLOS
LOS fading parameter for each multi-path component	$m = 5$
NLOS fading parameter of each multi-path component	$m = 1$
Number of multi-path signals	$L = 5$
Number of users	$K = 10$
Number of RAKE fingers	$L_R = 2$
Number of beamforming elements	$M_B = 3$

Table 6.2. System parameters for numerical evaluation of BER performance with combined beamforming and diversity.

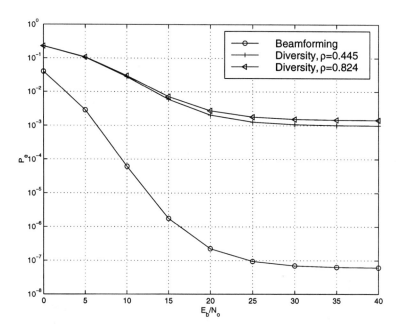

Figure 6.16. A comparison of $M_D = 3$ antenna diversity and $M_B = 3$ element beamforming systems under LOS conditions.

correlation between the fading envelopes of signals received at the various diversity branches will increase the BER. Turning our attention to the beamforming performance, the BER results obtained with $M_B = 3$ antennas are significantly better than with a $M_D = 3$ antenna diversity system. This feature is explained in detail below.

Figure 6.17 shows a similar comparison between diversity and beamforming systems under NLOS conditions. As would be expected, the absolute BER performance is worse as a result of the fading that is more severe, however, the general trends of correlation negatively influencing diversity performance and beamforming yielding better results than diversity are continued.

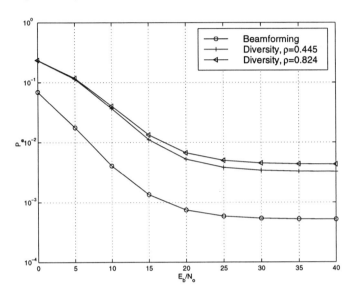

Figure 6.17. A comparison of $M_D = 3$ antenna diversity and $M_B = 3$ element beamforming systems under NLOS conditions.

A comparison between a beamforming system and a combined beamforming and diversity system under LOS conditions is shown in Figure 6.18. In this case, both system utilize nine antenna elements. The beamformer uses a $M_B = 9$ element ULA, with the combined system using three, $M_B = 3$ element ULAs. The first significant point to notice is that the combined system is more sensitive to non-zero correlation values as opposed to the standard diversity system discussed above. Secondly, the beamformer again performs better than the combined system. Under NLOS conditions the picture changes dramatically. In this case, the combined beamforming and diversity scheme outperforms the beamformer substantially (Figure 6.19). Also, the effect of correlation can still be seen.

Based on the results presented above, the following general conclusions can be made:

- Non-zero correlation values significantly influence the performance of a diversity system.

- The effect of non-zero correlation becomes less noticeable when the MAI present is high.

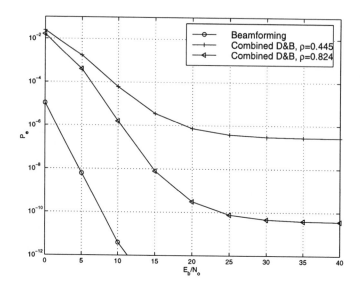

Figure 6.18. A comparison of a combined $M_B = 3$ element antenna beamformer with $M_D = 3$ diversity branches and a $M_B = 9$ element beamforming systems under LOS conditions.

- When the levels of MAI is high, beamforming yields better BER results than diversity.

These results may seem somewhat surprising, but are in line with established theory. It is well known and also shown in Section 5.2.2, that diversity systems provide no performance gain in AWGN environments. Furthermore, when the levels of MAI in a CDMA system is high it is the dominating effect in the BER performance and the channel starts to approximate a Gaussian channel (viz. the standard Gaussian assumption). Therefore, the effectivity of the diversity system reduces. On the other hand, the beamforming system (or even a system with sectorization) will always reduce the levels of interference by "removing" a number of users from the system resulting in better BER performance. The contrary is also true. When the number of users are low, and the channel is severely fading (as in the NLOS case), the beamforming system cannot improve the receive signal as it does not add any new information to the received signal, it merely limits the MAI. Thus, beamforming systems do not achieve high performance gains. On the other hand, a diversity system combining several severely fading signals will lead to much improved system performance, especially when the fading effects are dominant over the MAI effects.

Based on the discussion above, the following guidelines for the choice between diversity, beamforming and combinations thereof can be made:

- In high user, low fading environments, beamforming will perform best.

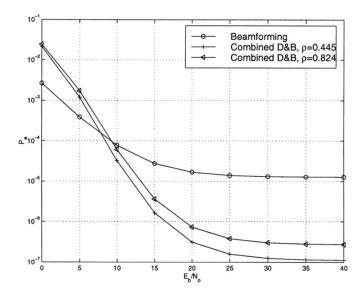

Figure 6.19. A comparison of a combined $M_B = 3$ element antenna beamformer with three diversity branches and a $M_B = 9$ element beamforming systems under NLOS conditions.

- In low user, severe fading environments diversity will perform best.

- In high user, severe fading environments, a combined diversity and beamforming approach will perform best.

- In low user, low fading environments, the performance of the systems will be approximately equal.

6.5 MULTIUSER MODULATION SCHEMES

In this section it will be shown that the conventional wisdom of single user modulation schemes does not necessarily hold in a multiple access environment. This fact has been pointed out first in [34] for convolutionally coded BPSK and QPSK multiple access systems. To illustrate the point in more detail, we will discuss the difference between BPSK and QPSK modulation under realistic multiple access situations.

It is well known that higher order modulation schemes allow for more efficient signalling, such as the relationship between BPSK and QPSK modulation in which QPSK offers twice the bandwidth efficiency of BPSK without sacrificing the average error rate as a function of the E_b/N_0. In a typical mobile multiple access communication channel, the multiple access interference experienced by any user in a CDMA system will be complex-valued due to independent phase offsets between signals received from different users. In ad-

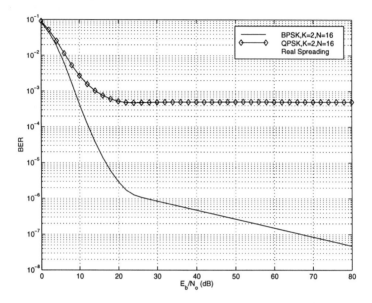

Figure 6.20. Probability of error in two-user BPSK and QPSK system with $N = 16$.

dition, complex spreading codes may also be employed, in which case the MAI is complex-valued, even without phase offsets. In [151] it has been pointed out that the single user relationship between BPSK and QPSK, described above, does not hold in the presence of complex MAI, which arises when

- real-valued spreading codes are used in a fading environment, resulting in phase offsets between users' signals, or
- complex-valued spreading codes are used, regardless of the environment fading.

This difference in performance can be attributed to the fact that only the real part of the MAI is significant for BPSK demodulation, whereas both real and imaginary components are used in QPSK demodulation. The lower MAI experienced by the BPSK receiver for a given number of users results in lower error rates, and immediately reveals one significant difference between single- and multiuser BPSK and QPSK systems. For useful comparison between systems, we define bandwidth efficiency as

$$\eta_m = \frac{K \cdot \mathcal{R}}{\mathcal{BW}} = \begin{cases} \alpha'_m \frac{K}{N} \text{ for BPSK} \\ \alpha'_m \frac{2 \cdot K}{N} \text{ for QPSK} \end{cases}, \quad (6.26)$$

where, as before, $\mathcal{R}$ is the bit rate, $\mathcal{BW}$ is the total bandwidth occupied, K is the number of users, N the processing gain and α'_m is a proportionality

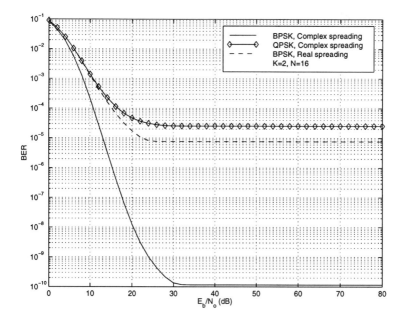

Figure 6.21. Probability of error in two-user BPSK and QPSK system with complex spreading sequences and $N = 16$.

constant. As shown in [151], for the same η_m, the performance of BPSK and QPSK systems are very similar in the presence of complex MAI.

To illustrate the discussion above, results from [151, 152] are illustrated below. For a very simple two user ($K = 2$) case with $N = 16$, the average BER for BPSK and QPSK is illustrated in Figures 6.20 and 6.21 for real and complex valued spreading codes respectively. From Figure 6.20, the BPSK system performs much better that the QPSK system, although the bandwidth efficiency, as defined in (6.26), of the former is only half of the latter. Another important point to note is also that, for the QPSK system, the error floor is much higher than for the BPSK system, which is an indication of more MAI being present. A similar trend can be observed for complex spreading sequences in Figure 6.21. It should be noted that the (very low) error observed for the complex valued BPSK result is due to the zero phase offset simulation environment, which is realistic for the downlink of a cellular system, in which the mobile user receives all other users' signals over the same base station-to-mobile channel.

Figure 6.22 summarizes our main conclusions, in that with regard to the performance for a conventional matched filter receiver, the following systems have the same performance: BPSK(K, N), QPSK($K, 2 \cdot N$) and QPSK($\frac{1}{2}K, N$). When interference canceling receivers (described in more detail in Chapter 7) are used, the results appear to hold approximately as indicated in Figure 6.23

for a successive interference canceler (SIC). With the decorrelator detector, however, the single-user result that QPSK always has better bandwidth efficiency than BPSK holds.

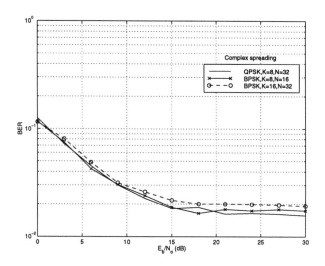

Figure 6.22. Probability of error performance for three equivalent systems of the same bandwidth efficiency, using complex spreading codes and assuming independent one-path fading channels for each user.

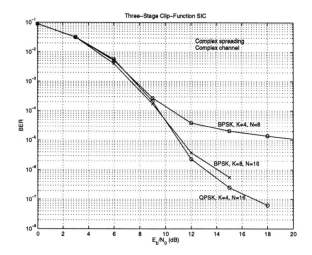

Figure 6.23. Probability of error performance over all users at the output of the third stage of a clip-function SIC, for $K = 4, 8$ BPSK and $K = 4$ QPSK systems having identical bandwidth efficiencies (complex spreading codes and complex channels).

From the above discussion, it is important to note that conventional single-user wisdom does not necessarily hold in a MU-MIMO system. This discussion also allows us to draw conclusions from space-time systems when only BPSK (or *vice versa*) results are available. Therefore, the average error rate results derived for space-time systems employing BPSK modulation in the preceding sections can be easily extended to QPSK based systems such as UMTS.

6.6 SUMMARY

In this chapter, the BER performance of various space-time processors has been derived analytically and some numerical results were presented. Based on the analytic derivations, the close relationship between diversity and beamforming systems have been shown and basic guidelines for choosing between diversity, beamforming and combined diversity and beamforming systems were discussed. The performance evaluation techniques discussed in this chapter will be extended in the next two chapters to include multiuser detection and space-time coding techniques.

7 MULTIUSER DETECTION

In this chapter, we consider the use of multiuser detection in conjunction with antenna arrays. Multiuser detection is an area of increasing importance in CDMA and has now reached a certain level of maturity. Here, we attempt to provide an overview of the underlying concepts and principles behind multiuser detection. The most fundamental detector structures will be presented and discussed to highlight design principles and potential advantages and disadvantages, as well as to relate different structures to one another. As the area of multiuser detection is rich in detail and approaches, the presentation here will be kept brief in order to attain the objective of providing an overview. For more detailed presentation and discussion, the reader is referred to [302] and the references found throughout this chapter.

In the discussion, the use of antenna arrays in conjunction with multiuser detection will be considered. As the objective of multiuser detection is to resolve MAI, a potential antenna array is most effectively used to provide maximum output SNR rather than to directly limit interference through null steering. Assuming a perfect multiuser detector that completely resolves MAI, optimal performance is only achievable if maximum SNR is guaranteed. The antenna beam should therefore be pointed directly at the angular location of the desired user, regardless of interferer locations. This is in line with the approach considered in previous chapters.

This chapter was written by L. Rasmussen, Chalmers University of Technology, Sweden

Using this approach, the antenna array provides extra dimensionality to the detection process where in fact it is possible to show that an M_D element array on average extends the effective processing gain by almost a factor of M_D [219]. In some sense, the antenna array represents a spatial RAKE combiner. Another advantage of this approach is that the resulting channel model is identical in form to the model most effectively used for multiuser detection design [216]. The antenna array is then naturally incorporated into the detector structure as we will see in the following.

7.1 MULTIUSER DETECTION

For almost a decade, multiuser detection for CDMA was considered to be prohibitively complex for practical implementation. However, with the massive research conducted in conjunction with UMTS, this view has now changed.

A major breakthrough came in the mid eighties when researchers such as Kohno [130], and Verdú [297, 298] showed that the MAI limitation was not inherently part of CDMA modulation, but a consequence of attempting to use the correlator receiver. In [297] Verdú proposed the optimum multiuser CDMA receiver based on a bank of matched filters followed by a maximum-likelihood sequence detector (MLSD). The objective of the MLSD is to find the input sequence which maximizes the conditional probability, or likelihood of the given output sequence. For asynchronous CDMA (where user transmissions are not co-ordinated) the MLSD may be implemented using the Viterbi algorithm [303]. This will be discussed in more detail later on.

As may be expected, however, a MLSD receiver requires extensive computational complexity which grows exponentially with the number of users in the system. It is, therefore, not feasible for implementation. It did, however, open the way for a flurry of research on suboptimal, lower complexity receiver structures which allow the benefits of CDMA to be realized.

Multiuser detectors can be divided into two main classes; one-shot multiuser detectors and interference cancellation (IC), or iterative multiuser detectors. In previous literature, one-shot receivers have usually been referred to as joint multiuser detectors as opposed to interference cancellation detectors. This terminology is somewhat unfortunate as both classes are indeed performing joint multiuser detection. In this chapter, we will therefore refer to detectors where each user is processed only once before final decisions are made, as one-shot detectors. Structures where an explicit estimate of the MAI component in the received signal is generated and then subtracted from the received signal in an iterative manner will be denoted interference cancellation schemes.

The remainder of the chapter is organized as follows. We first extend the model presented in Chapter 2, to a discrete-time model of the communication system based on linear algebra. This model is particularly suitable for detection design. An overview of the class of one-shot detectors is then presented where the conventional correlation detector, the optimal ML detector and the optimal linear ML detector are discussed in some detail. The MMSE detector is also considered with a brief discussion of possible adaptive implementation. The focus is then directed towards interference cancellation where linear cancellation is presented in some detail, followed by discussions on non-linear approaches. The theoretical part of the chapter is completed by a brief discussion on joint decoding for coded CDMA.

As interference cancellation schemes are most interesting for practical implementation, a series of simulation results are presented which demonstrate the BER performance of various approaches. The impact of antenna arrays on the BER performance is also illustrated by examples.

7.2 SYSTEM MODEL FOR DETECTION

In this section the system model derived in Chapter 2 is expanded to include a compact notation for the multiuser detection decision statistic. The system model derived here will form the basis for the rest of the discussion on multiuser detection.

Assuming that the desired signal is multi-path component q of subscriber j, the output of the receiver for symbol interval i after correlation, demodulation and sampling is given by (2.16). For multiuser detection over $P+1$ bits, the decision statistic of (2.16) can be written as

$$\zeta = \left(\zeta_1^{(1)}(0), \zeta_2^{(1)}(0), \cdots, \zeta_L^{(1)}(0), \zeta_1^{(2)}(0), \cdots, \zeta_l^{(k)}(i), \cdots, \zeta_L^{(K)}(P)\right)^\top, \quad (7.1)$$

a length $(P+1)KL$ column vector. The vector of decision statistics can then algebraically be described as

$$\zeta = \mathbf{R}_\zeta \mathbf{C} \mathbf{b} + \eta_\zeta \qquad \in \mathbb{C}^{(P+1)KL}, \quad (7.2)$$

where

$$\mathbf{b} = \left(b_0^{(1)}, b_0^{(2)}, \cdots, b_0^{(K)}, b_1^{(1)}, \cdots, b_P^{(K)}\right)^\top \in \{-1, 1\}^{(P+1)K}, \quad (7.3)$$

$$\eta_\zeta = \left(\eta_1^{(1)}(0), \eta_2^{(1)}(0), \cdots, \eta_L^{(1)}(0), \cdots, \eta_l^{(k)}(i), \cdots, \eta_L^{(K)}(P)\right)^\top$$
$$\in \mathbb{C}^{(P+1)KL}, \quad (7.4)$$

$$\mathbf{C} = \mathrm{diag}(\mathbf{C}_0, \mathbf{C}_1, \cdots, \mathbf{C}_P) \in \mathbb{C}^{(P+1)KL \times (P+1)K}, \quad (7.5)$$

$$\mathbf{R}_\zeta = \begin{bmatrix} \mathbf{R}_0^{(0)} & \mathbf{R}_0^{(1)} & & & & \\ \mathbf{R}_1^{(-1)} & \mathbf{R}_1^{(0)} & \mathbf{R}_1^{(1)} & & \mathbf{O} & \\ & \mathbf{R}_2^{(-1)} & \mathbf{R}_2^{(0)} & \mathbf{R}_2^{(1)} & & \\ & & & \ddots & & \\ & \mathbf{O} & & \mathbf{R}_{P-1}^{(-1)} & \mathbf{R}_{P-1}^{(0)} & \mathbf{R}_{P-1}^{(1)} \\ & & & & \mathbf{R}_P^{(-1)} & \mathbf{R}_P^{(0)} \end{bmatrix}$$

$$\in \mathbb{C}^{(P+1)KL \times (P+1)KL},$$

$$\mathbf{R}_i^{(1)} = \left(\mathbf{R}_{i+1}^{(-1)}\right)^\mathrm{H},$$

and

$$\mathbf{C}_i = \operatorname{diag}(\mathbf{c}_i^{(1)}, \mathbf{c}_i^{(2)}, \cdots, \mathbf{c}_i^{(K)}) \in \mathbb{C}^{KL \times K}, \tag{7.6}$$

$$\mathbf{c}_i^{(k)} = \left(c_1^{(k)}(i), c_2^{(k)}(i), \cdots, c_L^{(k)}(i)\right)^\mathrm{T} \in \mathbb{C}^L, \tag{7.7}$$

$$\mathbf{R}_i^{(x)} = \begin{bmatrix} \rho_{11,11}^{(x)}(i) & \rho_{11,12}^{(x)}(i) & \cdots & \rho_{11,kl}^{(x)}(i) & \cdots & \rho_{11,KL}^{(x)}(i) \\ \rho_{12,11}^{(x)}(i) & \rho_{12,12}^{(x)}(i) & & \rho_{12,kl}^{(x)}(i) & \cdots & \rho_{12,KL}^{(x)}(i) \\ \vdots & & & & & \vdots \\ \rho_{kl,11}^{(x)}(i) & \rho_{kl,12}^{(x)}(i) & & \rho_{kl,kl}^{(x)}(i) & \cdots & \rho_{kl,KL}^{(x)}(i) \\ \vdots & & & & & \vdots \\ \rho_{KL,11}^{(x)}(i) & \rho_{KL,12}^{(x)}(i) & \cdots & \rho_{KL,kl}^{(x)}(i) & \cdots & \rho_{KL,KL}^{(x)}(i) \end{bmatrix},$$

with

$$\rho_{jq,kl}^{(0)}(i) = \frac{A_k}{2} \exp\left(-j\left(\theta^{(j)} - \theta^{(k)}\right)\right) \|\mathbf{w}^{(k)}\| \overline{\mathcal{R}}^{(jk)} R_{jq,kl}^{(0)}(i), \tag{7.8}$$

$$\rho_{jq,kl}^{(-1)}(i) = R_{jq,kl}^{(-1)}(i), \tag{7.9}$$

$$\rho_{jq,kl}^{(1)}(i) = R_{jq,kl}^{(1)}(i). \tag{7.10}$$

The noise vector η_ζ is a zero-mean, complex multivariate Gaussian distribution with covariance matrix,

$$\mathrm{E}\{\eta_\zeta \eta_\zeta^\mathrm{H}\} = \mathbf{R}_\zeta \sigma^2. \tag{7.11}$$

The decision statistic vector ζ provides one decision statistic per multi-path component.

The channel model in (7.2) includes all the mobile radio parameters considered in this book and exhibits the same general algebraic structure that almost all multiuser detection strategies are based on[1]. It is therefore conceptually straightforward to apply results available in the literature to this more practical scenario.

7.3 ONE-SHOT DETECTION TECHNIQUES

7.3.1 Conventional Detection

Conventional detection in a single path transmission environment is done by matched filtering and sampling of the received signal, followed by a decision device, e.g., a simple polarity check for BPSK. The received signal is matched to the spreading code of the desired user as shown in (2.16). In a single user environment, this is optimal in the sense that the SNR is maximized which in turn corresponds to maximum-likelihood (ML) detection [205]. In a multiuser environment, this is, however, not entirely true. The SNR is still maximized, but the detector is not ML due to the presence of MAI.

In a multi-path environment, we can obtain decision statistics for each multi-path component as shown in (7.2). There are then several strategies for receive diversity[2] combining of these decision statistics pertaining to the same bit. The three classical strategies are selection diversity, equal gain combining and MRC. Combining can, of course, also be done after individual detection of each multi-path component which does in fact provide potentially better performance [219]. This is, however, not as common as pre-detection combining. We will therefore only consider pre-detection diversity combining.

The combined decision statistic vector $\mathbf{y}_c$, consists of one decision variable per user per bit, so

$$\mathbf{y}_c = (\mathbf{G}_c)^H \zeta \in \mathbb{C}^{(P+1)K}, \qquad (7.12)$$

where $\mathbf{G}_c$ is determined according to the combining strategy of choice. For selection diversity,

$$\mathbf{G}_s = \mathrm{diag}(\mathbf{G}_0^{(s)}, \mathbf{G}_1^{(s)}, \cdots, \mathbf{G}_P^{(s)}) \in \mathbb{C}^{(P+1)KL \times (P+1)K}, \qquad (7.13)$$

$$\mathbf{G}_i^{(s)} = \mathrm{diag}(\mathbf{g}_s^{(1)}(i), \mathbf{g}_s^{(2)}(i), \cdots, \mathbf{g}_s^{(K)}(i)) \in \mathbb{C}^{KL \times K}, \qquad (7.14)$$

$$\mathbf{g}_s^{(k)}(i) = \left(0, 0, \cdots, 0, \exp(j\varphi_s^{(k)}(i)), 0, \cdots, 0\right)^T \in \mathbb{C}^L, \qquad (7.15)$$

where the non-zero position selects the strongest multi-path component for user k at symbol interval i, i.e., the position of the component of $\mathbf{c}_i^{(k)}$ which has the largest value of $\beta_l^{(k)}(i)$.

For equal gain combining, the matrix combining filter $\mathbf{G}_e$, has the same general structure as in (7.13) and (7.14). In this case, however,

$$\mathbf{g}_e^{(k)}(i) = \left(\exp(j\varphi_1^{(k)}(i)), \exp(j\varphi_2^{(k)}(i)), 0, \cdots, 0, \exp(j\varphi_q^{(k)}(i)), 0, \cdots, 0\right)^T$$
$$\in \mathbb{C}^L,$$

where the three strongest paths are used for combining[3].

The conceptual equivalent to MRC, usually termed RAKE reception[4], can be described through a matrix combining filter $\mathbf{G}_R$, with the same structure as in (7.13) and (7.14) with

$$\mathbf{g}_R^{(k)}(i) = \left(c_1^{(k)}(i), c_2^{(k)}(i), 0, \cdots, 0, c_q^{(k)}(i), 0, \cdots, 0\right)^T \in \mathbb{C}^L. \qquad (7.16)$$

Again, we have assumed that the 3 strongest paths are used for combining, i.e., a 3-finger RAKE is used.

Of these three different resulting decision statistics, equal gain combining is rarely used. Most common is either selection diversity or RAKE combining. In the remainder of this chapter, we will assume that RAKE combining is performed at the front-end of the multiuser detector. The corresponding decision statistic is then

$$\mathbf{y} = (\mathbf{G}_R)^H \mathbf{R}_\zeta \mathbf{C} \mathbf{b} + (\mathbf{G}_R)^H \boldsymbol{\eta}_\zeta = \mathbf{R} \mathbf{b} + \boldsymbol{\eta}, \qquad (7.17)$$

where $\mathbf{R} = (\mathbf{G}_R)^H \mathbf{R}_\zeta \mathbf{C}$ and $\boldsymbol{\eta} = (\mathbf{G}_R)^H \boldsymbol{\eta}_\zeta$. $\mathbf{R}$ has the same general structure as $\mathbf{R}_\zeta$.

From (7.17) it is clear that several signals **b**, are fed into the channel and several signals emerge from the channel **y**. When we form **y** from $\mathbf{r}(t)$ by matched filtering we get precisely one output for every input.

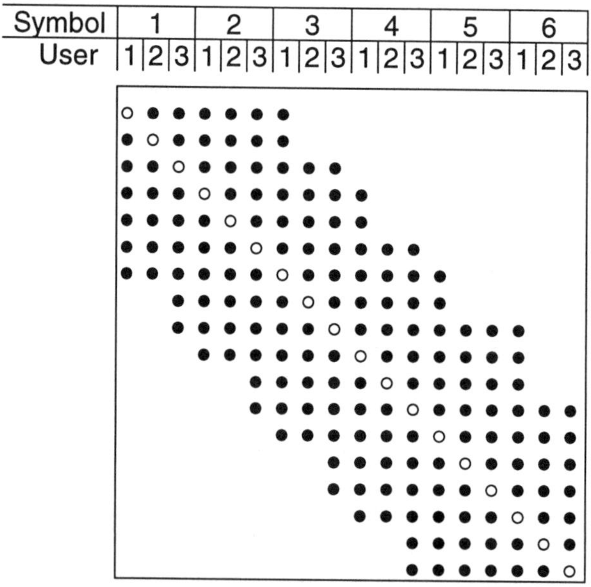

Figure 7.1. The fundamental structure of **R** where $K = 3$ and $P = 5$. The circles on the diagonal represent the autocorrelations. The remaining dots represent non-zero cross correlations. The band-diagonal structure is obvious.

Interference arises since, in general, every output has a contribution from every input. Under ideal conditions of orthogonal spreading codes there would be no interference, but such a scenario is virtually impossible in practice since we would require an improbable coincidence of multi-path channel, spreading code and DOA. In the general case where any output $y_i^{(k)}$ potentially depends on every input $\{b_i^{(k)} : i = 0, \cdots, P, k = 1, \cdots, K\}$, we need not even make the direct association between $y_i^{(k)}$ and $b_i^{(k)}$. Indeed, in pathological cases, some other observation, $y_n^{(j)}$ say, may contain more information about $b_i^{(k)}$ than $y_i^{(k)}$. In practice, however, we find that $b_i^{(k)}$ is among the strongest contributors to $y_i^{(k)}$ and that the other contributors are localized as indicated by the band-diagonal structure of **R** illustrated in Figure 7.1. Most of the interference in $y_i^{(k)}$ comes from symbols transmitted within a symbol interval or two of the symbol intervals.

In order to provide a feeling for the components constituting each observation $y_i^{(k)}$ we consider the components of (7.17) pertaining to $y_i^{(k)}$. For the noiseless case, we can write,

$$\mathbf{y} = \mathbf{R}\mathbf{b} = (\mathbf{D} + \mathbf{M})\mathbf{b}, \tag{7.18}$$

where **D** is a diagonal matrix with the diagonal elements of **R**. Then for $y_i^{(k)}$

$$y_i^{(k)} = d_i^{(k)} b_i^{(k)} + \left(\mathbf{m}_i^{(k)}\right)^H \mathbf{b},$$

where $\left(\mathbf{m}_i^{(k)}\right)^H$ is the row pertaining to symbol interval i, user k of matrix **M** and $d_i^{(k)}$ is the corresponding diagonal element of **D**. It is of interest to compare the average size of the two components, one being the signal component and the other being the interference component. Assuming perfect knowledge of the channel, and, therefore, perfect matched filtering, the power in the interference term is shown in Figure 7.2 as a function of K.

The desired signal component is the horizontal line at 1 for $d_i^{(k)} = 1$. We see that the interference power should not be ignored and, moreover, should be removed as effectively as possible. If not, effective signal to interference ratios well below 0 dB are possible, even following the inherent enhancements provided by the processing gain. It is apparent from Figure 7.2 that if the interference is treated as random noise that cannot be estimated then the channel capacity per user will go to zero as the number of users increases [299]. Although this is also true when the structure of the interference is exploited through joint detection, the degradation is more rapid in the former case. It is true that the CDMA channel is in some sense interference limited[5], but this limit is imposed by the adder channel nature of CDMA, not by excessive noise in correlator outputs.

MAI can be fairly accurately approximated as Gaussian noise when the number of users is large [206]; hence for some time the conventional detector was considered optimal and CDMA as a technology was deemed strongly interference limited. With the emergence of multiuser detection, however, where the particular structure of the MAI is being exploited, it is now well-known that the conventional detector is suboptimal. It is therefore the conventional detector which is highly interference limited and in general not CDMA as a technology.

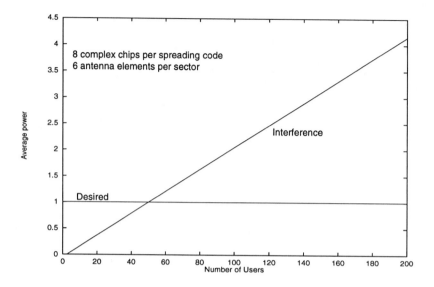

Figure 7.2. Interference power versus desired signal power as a function of the number of active users.

The conventional detector still plays an important role in multiuser detection since the output of a bank of matched filters provides a minimal sufficient statistic for detection [297]. Most multiuser detectors therefore include a conventional matched filter bank as the receiver front-end, followed by a RAKE combiner.

7.3.2 Optimal Detection

The Optimal detector is defined as follows,

$$\hat{\mathbf{b}}_{ML} = \arg\max_{\mathbf{b} \in \{-1,1\}^{(P+1)K}} P(\mathbf{b}|\mathbf{y}) \qquad (7.19)$$

$$= \arg\max_{\mathbf{b} \in \{-1,1\}^{(P+1)K}} P(\mathbf{b}|\mathbf{y})p(\mathbf{y}) \qquad (7.20)$$

$$= \arg\max_{\mathbf{b} \in \{-1,1\}^{(P+1)K}} p(\mathbf{y}|\mathbf{b})P(\mathbf{b}) \qquad (7.21)$$

$$= \arg\max_{\mathbf{b} \in \{-1,1\}^{(P+1)K}} p(\mathbf{y}|\mathbf{b}), \qquad (7.22)$$

where "arg max" denotes the data vector (argument) $\mathbf{b} \in \{-1,1\}^{(P+1)K}$ that maximizes the pdf of receiving vector $\mathbf{y}$ given that vector $\mathbf{b}$ was transmitted and that $P(\mathbf{b}) = 2^{-(P+1)K}$, i.e., all data vectors are independent and equally likely or equivalently $P(b_i^{(k)} = 1) = P(b_i^{(k)} = -1) = \frac{1}{2}$ for all i and k. This is what is also known as the MLSD.

Assuming that all the spreading codes are known and given knowledge of **b**, then the only random component in **y** is the Gaussian noise η. It therefore follows that **y** is Gaussian distributed with $E\{y\} = Rb$ and $\mathrm{var}\{y\} = \sigma^2 R$. The corresponding pdf is thus [215]

$$p(\mathbf{y} \mid \mathbf{b}) = K_0 \exp\left(-\frac{(\mathbf{y} - \mathbf{Rb})^H \mathbf{R}^{-1}(\mathbf{y} - \mathbf{Rb})}{2\sigma^2}\right), \qquad (7.23)$$

where K_0 is a constant independent of **y** and **b**, and therefore,

$$\begin{aligned}
\hat{\mathbf{b}}_{ML} &= \arg \min_{\mathbf{b} \in \{-1,1\}^{(P+1)K}} (\mathbf{y} - \mathbf{Rb})^H \mathbf{R}^{-1}(\mathbf{y} - \mathbf{Rb}) \\
&= \arg \min_{\mathbf{b} \in \{-1,1\}^{(P+1)K}} \mathbf{b}^T \mathbf{R} \mathbf{b} - 2\mathrm{Re}\{\mathbf{b}^T \mathbf{y}\}. \qquad (7.24)
\end{aligned}$$

Each possible data vector $\hat{\mathbf{b}}$, can be associated with a likelihood metric,

$$\lambda(\hat{\mathbf{b}}) = \hat{\mathbf{b}}^T \mathbf{R} \hat{\mathbf{b}} - 2\mathrm{Re}\{\hat{\mathbf{b}}^T \mathbf{y}\}.$$

We can then also write (7.24) as

$$\hat{\mathbf{b}}_{ML} = \arg \min_{\mathbf{b} \in \{-1,1\}^{(P+1)K}} \lambda(\mathbf{b}). \qquad (7.25)$$

The solution to the ML problem was proposed by Verdú in [297]. As the allowable alphabet is a discrete, finite set, the problem is equivalent to a combinatorial quadratic minimization and it is classified as an NP-hard problem [301]. In other words, there is no known algorithm that can solve the problem in polynomial time[6]. Only an exhaustive search through all $2^{(P+1)K}$ possible data vectors will guarantee a global solution. It is hence clear that the problem grows exponentially in complexity with the number of users. It also appears that the complexity grows exponentially with the sequence length. This is, however, not true. In the decision statistic **y**, described by (7.17), **R** is block diagonal as indicated in Figure 7.1. This is the same structure as an encoding matrix representing a convolutional error control encoding process. We can therefore consider **R** as representing a rate one time-varying, convolutional encoding of the input stream **b** having at least 2^{K-1} states for binary modulation formats.

As a consequence, **y** can be interpreted as a noisy output of a convolutional encoder with constraint length $K - 1$. From error control coding, we know that such a finite state machine problem can be solved most effectively through dynamic programming where all possible data sequences can be represented as paths through a trellis [238]. It follows that the Viterbi algorithm can be applied for MLSD given an appropriate recursive metric.

A recursive metric can only depend on previously processed bits. In order to derive such a metric, partition **b** as follows,

$$\mathbf{b} = \left(\mathbf{b}_0^T, \mathbf{b}_1^T, \cdots, \mathbf{b}_P^T\right)^T \qquad (7.26)$$

$$\mathbf{b}_i = \left(b_i^{(1)}, b_i^{(2)}, \cdots, b_i^{(K)}\right)^T, \qquad (7.27)$$

162   SPACE-TIME PROCESSING FOR CDMA

and define
$$\tilde{\mathbf{b}}_i^{(k)} = \left(\mathbf{b}_0^T, \mathbf{b}_1^T, \cdots, \mathbf{b}_{i-1}^T, b_i^{(1)}, \ldots, b_i^{(k)}\right)^T. \tag{7.28}$$

The recursive metric, only dependent on the previously processed data symbols, first suggested by Verdú in [297] is then obtained as

$$\begin{aligned}\mu_i^{(k)}(\tilde{\mathbf{b}}_i^{(k)}) &= d_i^{(k)} b_i^{(k)} + \sum_{\substack{n=1 \\ n \neq k}}^{K} m_{i,i-1}^{(kn)} b_{i-1}^{(n)} b_i^{(k)} + 2 \sum_{j=1}^{k-1} m_{i,i}^{(kj)} b_i^{(j)} b_i^{(k)} \\ &\quad - 2\text{Re}\{y_i^{(k)} b_i^{(k)}\},\end{aligned} \tag{7.29}$$

where $m_{i,l}^{(kj)}$ is the element pertaining to the interference from user j, symbol interval l to user k, symbol interval i of $\mathbf{M}$. Therefore we can recursively determine the log-likelihood metric as

$$\lambda(\mathbf{b}) = \sum_{i=0}^{P} \sum_{k=1}^{K} \mu_i^{(k)}(\tilde{\mathbf{b}}_i^{(k)}).$$

It is thus clear that the Viterbi algorithm can be applied and the complexity of the Optimal detector only grows exponentially with the number of users and not with the sequence length. The same arguments hold for the derivation of the MAP detection algorithm (see Appendix F for a description of the MAP algorithm) for the case of whitened noise.

7.3.3   Linear Detection

Receivers belonging to the linear class of one-shot detectors apply a linear transformation to the received vector representing the outputs of a bank of matched filters.

7.3.3.1   Decorrelating Detector. The linear ML detector can be readily derived based on (7.24) and is usually termed the decorrelating detector since all the MAI has been resolved at the expense of noise enhancement [150]. In this case, the output vector $\hat{\mathbf{b}}$ is allowed to take on any real value. It then follows from (7.24) that the output of the decorrelator is

$$\hat{\mathbf{b}}_{Dec} = \arg \min_{\mathbf{b} \in \mathbb{R}^{(P+1)K}} \mathbf{b}^T \mathbf{R} \mathbf{b} - 2\text{Re}\{\mathbf{b}^T \mathbf{y}\}.$$

This is an unconstrained, quadratic minimization and is solved by equating the derivative of the objective function to zero.

$$\frac{\partial(\mathbf{b}^T \mathbf{R} \mathbf{b} - 2\text{Re}\{\mathbf{b}^T \mathbf{y}\})}{\partial \mathbf{b}^T} = 2\mathbf{R}\mathbf{b} - 2\mathbf{y} = 0$$
$$\Downarrow$$
$$\mathbf{R}\hat{\mathbf{b}}_{Dec} = \mathbf{y} \Rightarrow \hat{\mathbf{b}}_{Dec} = \mathbf{R}^{-1}\mathbf{y}. \tag{7.30}$$

The output of the decorrelator is then

$$\hat{b}_{Dec} = \mathbf{R}^{-1}\mathbf{y} = \mathbf{R}^{-1}\mathbf{R}\mathbf{b} + \mathbf{R}^{-1}\boldsymbol{\eta} = \mathbf{b} + \hat{\boldsymbol{\eta}},$$

where $\mathrm{E}\left\{\hat{\boldsymbol{\eta}}\hat{\boldsymbol{\eta}}^T\right\} = \sigma^2 \mathbf{R}^{-1}$.

The calculation of the inverse of $\mathbf{R}$ is complex and difficult to perform in real-time due to its size of $(P+1)K \times (P+1)K$ and consequently the inherent detection delay of at least $P+1$ symbol intervals. One approach to limit the complexity involved in calculating the inverse is to apply a sliding processing window. If the window is slid one symbol at a time, and spans only the $2K-1$ data symbols that directly interfere with the data symbol in question, then we have the one-shot decorrelator suggested by Verdú in [300]. We can also detect more than one user at a time and thus let the window slide in blocks of users. The most common approach is to detect all users in symbol interval i by considering a window of size $2W+1$ where W past and future symbol intervals are included. Several authors have suggested such algorithms all of which are quite similar [317, 316, 9, 123].

Since a matrix inverse represents the solution to a set of linear equations, classic techniques for such problems can be applied. The most efficient techniques are the iterative methods, such as the Jacobi and the Gauss-Seidel iterations [91]. Again, however, we are faced with the problem of the size of $\mathbf{R}$, hence also in this approach a processing window can be applied to circumvent the problem.

The windowing can be avoided altogether by carefully applying appropriate iterative techniques. Due to the block diagonal structure of $\mathbf{R}$, the iterative updates can be done progressively as the update for each symbol interval only depends on the updates for the previous and the following intervals. The iterative solution is then a direct approximation to the inverse of $\mathbf{R}$ and not to a windowed version of $\mathbf{R}$. This approach has been recognized to be identical to linear interference cancellation and thus we will defer any further discussions until the section on interference cancellation. It should be noted, however, that this iterative implementation is the most effective technique for realizing the decorrelating detector for any asynchronous CDMA channel.

7.3.3.2 MMSE Detector.
The ML strategy is not the only filter design criterion that can be applied. The minimum mean square error (MMSE) criterion can also be used for detector design. In addition to its uses for beamforming described in Chapter 5, the linear MMSE (LMMSE) is a linear detector $\mathbf{H}^H$,

$$\hat{b}_{MMSE} = \mathbf{H}^H \mathbf{y} \in \mathbb{C}^{(P+1)K},$$

determined such that $J_{\text{tot}}(\mathbf{H}^H)$ defined below is minimized.

$$\begin{aligned} J_{\text{tot}}(\mathbf{H}^{\text{H}}) &= \mathrm{E}\left\{\|\hat{\mathbf{b}}_{MMSE} - \mathbf{b}\|^2\right\} \\ &= \mathrm{E}\{\|\mathbf{H}^{\text{H}}\mathbf{y} - \mathbf{b}\|^2\} \\ &= \mathrm{E}\{\mathbf{y}^{\text{H}}\mathbf{H}\mathbf{H}^{\text{H}}\mathbf{y}\} + \mathrm{E}\{\mathbf{b}^{\text{H}}\mathbf{b}\} - 2\mathrm{Re}\{\mathrm{E}\{\mathbf{b}^{\text{H}}\mathbf{H}^{\text{H}}\mathbf{y}\}\}, \quad (7.31) \end{aligned}$$

The filter $\mathbf{H}^{\text{H}}$ that minimizes $J_{\text{tot}}(\mathbf{H}^{\text{H}})$ is obtained by setting the derivative of the right hand side of (7.31) with respect to $\mathbf{H}^{\text{H}}$ to zero, i.e.,

$$\frac{\partial e}{\partial \mathbf{H}^{\text{H}}} = \mathrm{E}\{2\mathbf{H}^{\text{H}}\mathbf{y}\mathbf{y}^{\text{H}} - 2\mathbf{b}\mathbf{y}^{\text{H}}\} = 0$$
$$\Rightarrow \mathbf{H}^{\text{H}} = \mathrm{E}\{\mathbf{b}\mathbf{y}^{\text{H}}\}\mathrm{E}\{\mathbf{y}\mathbf{y}^{\text{H}}\}^{-1},$$

where

$$\begin{aligned} \mathrm{E}\{\mathbf{b}\mathbf{y}^{\text{H}}\} &= \mathbf{R} \\ \mathrm{E}\{\mathbf{y}\mathbf{y}^{\text{H}}\}^{-1} &= \mathbf{R}^{-1}\left(\mathbf{R} + \sigma^2\mathbf{I}\right)^{-1}. \end{aligned}$$

The LMMSE filter is thus determined as [326]

$$\mathbf{H}^{\text{H}} = \left(\mathbf{R} + \sigma^2\mathbf{I}\right)^{-1},$$

which incidently is also the filter that maximizes the (SINR) at the detector output [169].

Thus, $\mathbf{y}$ is multiplied by a modified version of the inverse of the correlation matrix. The transformation balances the desire to eliminate MAI with the desire to minimize noise enhancement. The LMMSE detector generally performs better in terms of BER than the decorrelating detector since background noise is taken into account.

An important observation is that the original minimization of $J_{\text{tot}}(\mathbf{H}^{\text{H}})$ is equivalent to $(P+1)K$ decoupled minimization problems

$$\min_{\mathbf{h}_i^{(k)}} J(\mathbf{h}_i^{(k)}) = \min_{\mathbf{h}_i^{(k)}} \mathrm{E}\left\{\|\left(\mathbf{h}_i^{(k)}\right)^{\text{H}}\mathbf{y} - b_i^{(k)}\|^2\right\},$$

for $k = 1, \cdots, K$ and $i = 0, \cdots, P$. Each minimization problem is easily recognized as a classical, discrete-time Wiener filtering problem yielding the MMSE solution for each vector as [153]

$$\mathbf{h}_i^{(k)} = (0, \cdots, 0, 1, 0, \cdots, 0)^{\text{T}}\left(\mathbf{R} + \sigma^2\mathbf{I}\right)^{-1},$$

where the first vector basically picks out the column of the MMSE filter corresponding to user k for symbol interval i.

The above decoupling property of the MMSE receiver has important implications for receiver complexity. If only the kth user's transmitted bits are required to be demodulated and there is no need to detect any of the other bits transmitted by other interfering users, then only one transversal filter is required. However, the filter coefficients need to be recomputed for every bit interval. This observation readily generalizes to any subset of users, i.e., the detection of the set of users complementary to the desired set leaves the MMSE optimality of the desired set unchanged. Detection of such a subset of active users is relevant for instance in the uplink when out-of-cell interference is significant and must be taken into account, and on the downlink when only one signal is desired. The MSE for the desired users is unaffected by whether other users are being detected at the same time or not.

The complexity advantage of this decoupling is especially pronounced for adaptive implementations of the MMSE detector. Such structures are discussed in the following section.

In the direct implementation for an asynchronous system, we encounter the same problems with the MMSE detector as with the decorrelator. Again approximating techniques must be used as the required matrix inversion for the detector is generally too complex and introduces intolerable detection delays. The same solution techniques as discussed above for the decorrelator can also be applied to the MMSE detector, i.e., windowing techniques and iterative implementation, leading to interference cancellation structures. The iterative techniques will be discussed in more detail later on.

7.3.3.3 Adaptive MMSE Detection. The LMMSE receiver requires knowledge of the channel, the spreading codes and the noise variance. These requirements can be avoided by an adaptive realization of the detector. An alternative to a bank of conventional matched filters is a bank of adaptive filters, the outputs of which still represent soft estimates of the received data symbols for each user. The family of adaptive filters aims to minimize the error between the expected and the actual filter output. These filters can be implemented in a standard transversal FIR filter structure, with time-varying coefficients that are updated according to the error in the filter output. They require no information about the signals except for a known training sequence for the desired user and a rough timing estimate. They do, however, require the use of short spreading codes. Proposed detectors based on filtering the received signal include [169, 173, 208], while a comprehensive tutorial is presented in [153].

In general, when long codes are used, $\mathbf{h}_i^{(k)}$ is a function of the symbol index i and needs to be computed based on knowledge of the channel, the spreading codes of all the other users and the noise variance. Assuming that short codes[7] are used and that the channel only changes slowly with i, $\mathbf{h}_i^{(k)}$ will essentially be constant over a substantial number of symbol intervals. In this case, it becomes possible to iteratively search for $\mathbf{h}_i^{(k)}$ over many symbols using an adaptive algorithm operating at symbol rate to minimize the cost function

$$J(\hat{\mathbf{h}}_i^{(k)}) = \mathrm{E}\left\{\|\left(\hat{\mathbf{h}}_i^{(k)}\right)^{\mathrm{H}}\mathbf{y} - b_i^{(k)}\|^2\right\}, \qquad (7.32)$$

with respect to $\hat{\mathbf{h}}_i^{(k)}$. Unfortunately when long codes are used, $\mathbf{h}_i^{(k)}$ changes quite randomly from one symbol interval to the next, making it impossible for an adaptive algorithm (which can only track slow changes) to find the MMSE filter.

For an asynchronous system, we have seen above that optimal MMSE detection is accomplished only by using a window spanning all symbol intervals transmitted by all users. Clearly this is not feasible even for an adaptive detector and so it has also been suggested in this case that an observation window be used. In this case the adaptive MMSE filter $\hat{\mathbf{h}}_i^{(k)}$ will be of length $2W + 1$.

The steepest decent algorithm for the symbol-rate optimization of (7.32) can be used to iteratively find the filter vector, and is given by

$$\begin{aligned}\hat{\mathbf{h}}_{i+1}^{(k)} &= \hat{\mathbf{h}}_i^{(k)} - \mu \frac{\partial J(\hat{\mathbf{h}}_i^{(k)})}{\partial \hat{\mathbf{h}}_i^{(k)}} \\ &= \hat{\mathbf{h}}_i^{(k)} + \mu \mathrm{E}\left\{\left(e_i^{(k)}\right)^{\mathrm{H}} \mathbf{y}_i\right\},\end{aligned}$$

where $e_i^{(k)} = b_i^{(k)} - \left(\hat{\mathbf{h}}_{i+1}^{(k)}\right)^{\mathrm{H}} \mathbf{y}_i$ is the estimation error. The steepest decent algorithm may be approximated using instantaneous values in place of ensemble averages to yield the LMS updates

$$\hat{\mathbf{h}}_{i+1}^{(k)} = \hat{\mathbf{h}}_i^{(k)} + \mu \left(e_i^{(k)}\right)^{\mathrm{H}} \mathbf{y}_i.$$

The error sequence $e_i^{(k)}$ is generated using training symbols in the start-up mode and via decision feedback in a decision-directed mode of operation after transmission begins.

Since $\hat{\mathbf{h}}_i^{(k)}$ is updated only once per symbol interval, it is imperative that the optimum MMSE solution remains independent of i i.e., the optimum MMSE solution is the same over successive symbol intervals during training. This precludes having different spreading codes in different symbol intervals (or long codes) since such abrupt changes would disrupt the adaptation process.

As pointed out earlier on, the filter length ought to be infinite for optimal MMSE performance. However, for adaptive MMSE realizations, longer filters also imply increased misadjustment noise (which increases proportionately with filter length), while the additional gain in MMSE when an extra filter tap is added diminishes with filter length. Beyond a certain cross-over point where the increase in misadjustment over compensates for the decrease in MMSE by addition of more taps, increasing the filter length is clearly self-defeating. Issues of real-time computational complexity also preclude the use of filters that are too long. Filter order selection is therefore an issue of great practical importance that is solved through a judicious combination of design principles and experience gained via trial and error in practice.

As with all adaptation filters, the speed of convergence is of prime concern in an adaptive multiuser detector. It can be seen through simple computer simulations and calculations such as those presented in [153], that the convergence speed of the LMS algorithm for even moderately loaded systems is unsatisfactory. Training symbols of the order of at least hundreds are required. The RLS algorithm can be used in place of the LMS to significantly improve convergence. Increases in the convergence rate of around a factor of 5 can be expected [153] at the expense of a significant increase in computational complexity.

7.3.4 Decorrelating Decision Feedback Detector

The decorrelating decision feedback detector (DDFD) proposed by Duel-Hallen [58] uses both linear and nonlinear processing. The users are partially decorrelated by filtering with the partially decorrelating noise whitening (PDNW) filter. This filter is derived from a modified Cholesky decomposition of the correlation matrix, i.e., $\mathbf{R} = \mathbf{F}^H \mathbf{F}$ where $\mathbf{F}$ is lower left triangular. The corresponding filter is then $(\mathbf{F})^{-H}$, which will become clear below. The decision statistic that the DDFD works on is thus

$$\mathbf{x} = (\mathbf{F})^{-H}\mathbf{y} = (\mathbf{F})^{-H}(\mathbf{R}\mathbf{b} + \boldsymbol{\eta}) = \mathbf{F}\mathbf{b} + \hat{\boldsymbol{\eta}}, \qquad (7.33)$$

where $\mathrm{E}\left\{\hat{\boldsymbol{\eta}}\hat{\boldsymbol{\eta}}^\top\right\} = \sigma^2 \mathbf{I}$, i.e., the noise has been whitened. In addition, the effect of filtering with the PDNW filter is to partially decorrelate the users' signals. That is, the output of any individual user is only affected by MAI caused by previously processed data bits, and is not affected by MAI caused by any future data bits. The linear transformation is followed by a serial interference cancellation stage, where an estimate of the previously detected bits' MAI contribution is subtracted from the current symbol to be processed, and so on down to the last data symbol.

Since this scheme relies on a Cholesky decomposition and a matrix inverse, windowing techniques must again be applied [316, 9].

7.4 INTERFERENCE CANCELLATION TECHNIQUES

We have seen earlier on that each output symbol $y_i^{(k)}$ consists of a desired signal component and a MAI component and we have argued that the desired symbol predominantly determines the output symbol. Interfering symbols have, on average, a secondary influence on the output symbol. This fact can be used to great advantage in iterative detection strategies such as interference cancellation.

Although we know how to implement ML sequence detection, it is prohibitively complex, so perfect[8] interference cancellation is an attractive alternative since it efficiently removes all of the interference in each observation

$y_i^{(k)}$, leaving behind the desired contribution from $b_i^{(k)}$ in the presence of truly random noise. The problem with such an approach is that the receiver must know all of the interfering symbols. In practice, possibly erroneous estimates can be used instead. However, the quality of such estimates is paramount for resulting performance.

The first structure based on the principle of interference cancellation was the multi-stage detector in [294]. Here the cancellation is decision-directed (i.e., non-linear) and is done in parallel. In [53] Dent et al. proposed a serial approach, a single stage non-linear SIC scheme while Kawabe et al. suggested a multi-stage non-linear SIC technique in [125]. A closely related scheme was suggested by Sawahashi et al. in [235]. Linear SIC detectors have been considered in detail for both single- and multi-stage cases in [193] and [119], while Jamal and Dahlman have compared the performance of the linear and the non-linear SIC approaches in [116].

For practical implementation, cancellation structures working directly on the received chip-matched filtered statistics have mainly been suggested, e.g., [38, 234]. In this approach, a basic spreading/re-spreading building block is defined and used repeatedly to construct various cancellation schemes. Such an implementation, however, imposes certain constraints on the bit-level realization. For example, the chip-level S-stage parallel interference cancellation (PIC) experiences a detection delay of S symbol intervals, while a chip-level S-stage SIC encounters a delay of KS symbol intervals. If the canceler structure is implemented directly at bit-level, i.e., based on matched filtered outputs rather than chip-matched filtered outputs, then such constraints can be avoided at the expense of tighter constraints on processing speed.

In fact, for a bit-level implementation, all the algorithms considered here experience a detection delay of the order of $2S$, regardless of whether the fundamental cancellation process is successive or parallel. This is assuming that the correlation matrix, representing the transmission, is readily available. This assumption can be disputed, and to obtain the correlation matrix will almost certainly lead to additional delay. It will, however, be a constant delay, merely off-setting the quantified delays stated here.

The representation of the decision statistic in (7.18) is convenient for interference cancellation. Based on the matched filter output, we can make an estimate of the corresponding MAI as

$$I_{\text{MAI}} = \mathbf{M}\hat{\mathbf{b}}(0),$$

where $\hat{\mathbf{b}}(0) = \mathbf{f}_x(\mathbf{D}^{-1}\mathbf{y})$, $\mathbf{R} = \mathbf{D} + \mathbf{L} + \mathbf{U}$ where $\mathbf{L}$ is strictly lower triangular and $\mathbf{U}$ is strictly upper triangular and $\mathbf{M} = \mathbf{L} + \mathbf{U}$. Here, $\mathbf{f}_x(\cdot)$ is a tentative decision function to be discussed in more detail later on. The interference generated by all users is here estimated in parallel and hence the cancellation process is a parallel interference cancellation [294]. The decision statistic for the next stage $\mathbf{y}(1)$ is then obtained as

$$\mathbf{y}(1) = \mathbf{y} - \mathbf{M}\hat{\mathbf{b}}(0) = \mathbf{Db} + \mathbf{M}(\mathbf{b} - \hat{\mathbf{b}}(0)) + \eta,$$

and a new tentative decision is found as $\hat{\mathbf{b}}(1) = \mathbf{f}_x(\mathbf{D}^{-1}\mathbf{y}(1))$. In general the decision statistic and the tentative decision for stage s is

$$\begin{aligned}\mathbf{y}(s) &= \mathbf{y} - \mathbf{M}\hat{\mathbf{b}}(s-1), \\ \hat{\mathbf{b}}(s) &= \mathbf{f}_x(\mathbf{D}^{-1}\mathbf{y}(s)).\end{aligned}$$

The cancellation can also be done successively, i.e., one user is processed at a time where the most recent estimates of all other users are used to generate the interference estimates [193]. This can be described algebraically as

$$\begin{aligned}\mathbf{y}(s) &= \mathbf{y} - \mathbf{L}\hat{\mathbf{b}}(s) - \mathbf{U}\hat{\mathbf{b}}(s-1), \\ \hat{\mathbf{b}}(s) &= \mathbf{f}_x(\mathbf{D}^{-1}\mathbf{y}(s)),\end{aligned}$$

where it is noted that the elements of $\hat{\mathbf{b}}(s)$ and $\mathbf{y}(s)$ are determined successively.

The design of the decision function has significant influence on the nature and the performance of the various IC schemes. In the majority of the literature, either hard decisions or soft decisions (see Figure 7.3) have been used.

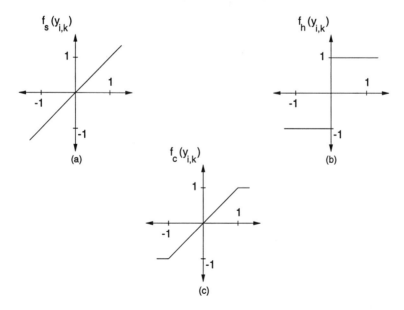

Figure 7.3. Different mapping functions, a) soft decision, b) hard decision, c) clipped soft decision.

Based on MMSE error considerations and convenient Gaussian approximations, it is possible to derive an optimal tentative decision function for the intermediate estimates [56]. For BPSK modulation formats, the function is a hyperbolic tangent of the matched filtered statistics $\mathbf{y}$, while for QPSK it is the same function applied independently to the in-phase and quadrature components. The piecewise linear clipped soft decision (CSD) (see Figure 7.3) is a good approximation of the hyperbolic tangent. In [190, 257], close to ML performance is reported for a multi-stage SIC based on the linear clipper.

The CSD function retains the advantages of both the linear and the hard-decision function while avoiding some of their obvious shortcomings. Intuitively, small correlator outputs lead to unreliable data estimates, and thus only a small amount of signal cancellation should be carried out. However, the hard limiter does not account for this effect and maps every correlator output onto one of the signal constellation points. On the other hand, large correlator outputs usually signify reliable decisions, and so full cancellation should be performed. With the linear mapping function, large outputs are not clipped but are wholly fed back into the detector, thereby resulting occasionally in large cancellation errors. The clip function circumvents these two problems, and therefore ought to produce better results.

Non-linear decision functions make analysis very difficult. Linear cancellation, however, can be described through linear algebra and as such provides insight into the mechanisms of cancellation. The linear cancellation structures do in fact, as pointed out earlier on, realize an iterative implementation of the optimal linear detectors. Due to the algebraic description and the intuitive insight the corresponding analysis provides into cancellation, we focus in some detail on linear cancellation.

7.4.1 Linear Interference Cancellation

Linear multiuser detectors process the matched filter output vector $\mathbf{y}$ in a one-shot linear operation. The detector output is $\mathbf{u}$ which satisfies

$$\mathbf{A}\mathbf{u} = \mathbf{y}. \tag{7.34}$$

For the decorrelating detector $\mathbf{A} = \mathbf{R}$ [167], and for the MMSE detector $\mathbf{A} = (\mathbf{R} + \sigma^2 \mathbf{I})$ [326], where the matrices $\mathbf{R}$ and $\mathbf{R} + \sigma^2 \mathbf{I}$ are symmetric positive definite (SPD), and block tridiagonal. Instead of inverting the matrix $\mathbf{A}$ using the direct method, we can solve the linear system (7.34) using iterative methods. This was first suggested in [124] for a windowed approach. Later it was realized in [61] that no windowing is required as the iterative structure in fact corresponds to multistage interference cancellation.

In [266], Tan and Rasmussen present a block iterative method which solves (7.34) in a timely manner without introducing additional complexity. The block iterative techniques are in fact identical to linear interference cancellation structures which will be made clear in the following.

In the bit-level representation of CDMA, a new set of decision statistics arrive every symbol interval, i.e., based on the inherent timing we can partition $\mathbf{y}$ into

$$\mathbf{y} = \left(\left(\mathbf{y}^{(0)}\right)^\mathsf{T}, \left(\mathbf{y}^{(1)}\right)^\mathsf{T}, \cdots, \left(\mathbf{y}^{(P)}\right)^\mathsf{T} \right)^\mathsf{T}.$$

This partitioning[9] also constitutes a natural grouping for block-wise cancellation. We therefore partition the problem in (7.34) into

$$\mathbf{A}\mathbf{u} = \begin{pmatrix} \mathbf{D}_{0,0} & \mathbf{U}_{0,1} & & & \\ \mathbf{L}_{1,0} & \mathbf{D}_{1,1} & \mathbf{U}_{1,2} & & \\ & \ddots & \ddots & \ddots & \\ & & \ddots & \ddots & \mathbf{U}_{P-1,P} \\ & & & \mathbf{L}_{P,P-1} & \mathbf{D}_{P,P} \end{pmatrix} \begin{pmatrix} \mathbf{u}^{(0)} \\ \mathbf{u}^{(1)} \\ \vdots \\ \vdots \\ \mathbf{u}^{(P)} \end{pmatrix} = \begin{pmatrix} \mathbf{y}^{(0)} \\ \mathbf{y}^{(1)} \\ \vdots \\ \vdots \\ \mathbf{y}^{(P)} \end{pmatrix} = \mathbf{y},$$

where $\mathbf{A} = \mathbf{D} + \mathbf{L} + \mathbf{U}$.

Consider a block iterative method which updates the users in blocks. Regarding each block as an "apparent" user, we essentially have a problem like (7.34) of dimension $(P+1)$ instead of dimension $(P+1)K$. Applying the Jacobi iteration to the block system, we get

$$\mathbf{D}_{n,n}\mathbf{u}_{m+1}^{(n)} = \mathbf{y}^{(n)} - \mathbf{L}_{n,n-1}\mathbf{u}_m^{(n-1)} - \mathbf{U}_{n,n+1}\mathbf{u}_m^{(n+1)},$$

while the Gauss-Seidel approach gives us

$$\mathbf{D}_{n,n}\mathbf{u}_{m+1}^{(n)} = \mathbf{y}^{(n)} - \mathbf{L}_{n,n-1}\mathbf{u}_{m+1}^{(n-1)} - \mathbf{U}_{n,n+1}\mathbf{u}_m^{(n+1)}. \qquad (7.35)$$

The block Gauss-Seidel is considered in the sequel since it is known to converge faster than the block Jacobi [99]. The block Gauss-Seidel then gets estimates of $\mathbf{D}_{n,n}\mathbf{u}_{m+1}^{(n)}$ by subtracting the most current estimates of the interference created by all the users in the previous symbol interval $\mathbf{L}_{n,n-1}\mathbf{u}_{m+1}^{(n-1)}$ and the next symbol interval $\mathbf{U}_{n,n+1}\mathbf{u}_m^{(n+1)}$ from the received signal $\mathbf{y}^{(n)}$. It is therefore clear that the block Gauss-Seidel in fact does block-wise linear SIC.

Equation (7.35) can also be viewed as a set of linear equations,

$$\mathbf{D}_{n,n}\mathbf{f}^{(n)} = \mathbf{v}^{(n)}. \qquad (7.36)$$

If (7.36) can be solved explicitly, then the block Gauss-Seidel is guaranteed to converge [99]. We will, however, also solve (7.36) iteratively in a so-called nested iteration [138] where the outer iteration is the block Gauss-Seidel and the inner iteration is selected to solve (7.36).

7.4.1.1 Jacobi Inner Iteration.
The simplest iteration method to consider for the inner iteration is the Jacobi iteration. It is represented by the

following splitting of the diagonal block matrix, $\mathbf{D}_{n,n} = \mathbf{D}_n + \mathbf{L}_n + \mathbf{U}_n$ with $\mathbf{D}_n$ being a diagonal matrix and $\mathbf{L}_n$ and $\mathbf{U}_n$ being strictly lower left and upper right triangular, respectively. If we consider the case where only one inner iteration is performed, then (7.36) becomes

$$\mathbf{D}_n \mathbf{u}_{m+1}^{(n)} = \mathbf{y}^{(n)} - \mathbf{L}_{n,n-1} \mathbf{u}_{m+1}^{(n-1)} - \mathbf{U}_{n,n+1} \mathbf{u}_m^{(n+1)} - (\mathbf{L}_n + \mathbf{U}_n) \mathbf{u}_m^{(n)}. \quad (7.37)$$

So, the stage $m+1$ estimates, $\mathbf{u}_{m+1}^{(n)}$ are obtained as the subtraction of the block interference estimates as discussed earlier on and the interference estimates from the inner iteration, i.e., m stage estimates $\mathbf{u}_m^{(n)}$. This constitutes a slightly modified linear PIC structure since stage $m+1$ estimates are used to generate the interference from the previous symbol interval. In case the block Jacobi iteration was used instead in (7.37) then it represents traditional parallel interference cancellation. However, the block Gauss-Seidel improves the performance at no additional expense.

It should be noted here that (7.37) does not require knowledge of the entire $\mathbf{A}$ and $\mathbf{u}_m$, only $\mathbf{D}_{n,n}$, $\mathbf{L}_{n,n-1}$ and $\mathbf{U}_{n,n+1}$ as well as $\mathbf{u}_m^{(n-1)}$, $\mathbf{u}_m^{(n)}$ and $\mathbf{u}_m^{(n+1)}$. Due to the sparse matrix structure, the Jacobi iteration decomposes into smaller problems described by (7.37) and hence, the equivalent IC scheme is recognized directly in the time domain as opposed to [61]. Furthermore, the block diagonal structure of $\mathbf{A}$ makes it possible to apply a cancellation structure without introducing inhibiting detection delays.

It should also be noted that the inner Jacobi iteration is not guaranteed to converge unless the spectral radius of $\mathbf{D}_n^{-1}(\mathbf{L}_n + \mathbf{U}_n)$ is less than one [99]. Convergence, however, is assured if the concept of over-relaxation is applied [91]. Hence, let us consider the following modification of the inner Jacobi step where the transition from $\mathbf{u}_m^{(n)}$ to $\mathbf{u}_{m+1}^{(n)}$ is given by

$$\begin{aligned}\mathbf{D}_n \mathbf{u}_{m+1}^{(n)} &= \mu_n \left(\mathbf{y}^{(n)} - \mathbf{L}_{n,n-1} \mathbf{u}_{m+1}^{(n-1)} - \mathbf{U}_{n,n+1} \mathbf{u}_m^{(n+1)} - (\mathbf{L}_n + \mathbf{U}_n) \mathbf{u}_m^{(n)} \right) \\ &\quad + (1 - \mu_n) \mathbf{D}_n \mathbf{u}_m^{(n)},\end{aligned}$$

This approach is known as the Jacobi over-relaxation (JOR) iteration and we will denote this block-wise JOR as B-JOR. It can easily be seen that the B-JOR is in fact a weighted linear PIC scheme where the interference estimate is scaled before subtraction. This way, a measure of reliability is introduced into the interference estimate. This type of weighted linear PIC structure is described in more detail in [56, 259, 98].

7.4.1.2 Gauss-Seidel Inner Iteration.

Another classic iterative method is the Gauss-Seidel iteration. Using the Gauss-Seidel method as the inner iteration, the transition from the tentative decision vector $\mathbf{u}_m^{(n)}$ to $\mathbf{u}_{m+1}^{(n)}$ is given by

$$(\mathbf{D}_n + \mathbf{L}_n) \mathbf{u}_{m+1}^{(n)} = \mathbf{y}^{(n)} - \mathbf{L}_{n,n-1} \mathbf{u}_{m+1}^{(n-1)} - \mathbf{U}_{n,n+1} \mathbf{u}_m^{(n+1)} - \mathbf{U}_n \mathbf{u}_m^{(n)}, \quad (7.38)$$

where again $\mathbf{D}_{n,n} = \mathbf{D}_n - \mathbf{L}_n - \mathbf{U}_n$. Solving (7.36) by a point Gauss-Seidel and (7.35) by a block Gauss-Seidel does in fact correspond to the ideal Gauss-Seidel. The block size, however, provides a degree of freedom to trade off detection delay for processing speed.

7.4.2 Non-linear Interference Cancellation

Non-linear cancellation has the potential to perform better than linear IC since there are no constricting linearity constraints. Traditionally, hard decisions have been used, and assuming that previous decisions for other users are correct, this corresponds to a ML decision. However, prior decisions are rarely all correct and thus, the performance is impaired. Other non-linear decision functions such as the CSD or the hyperbolic tangent can be used to improve performance. It remains true, however, that no theoretical approach for analyzing non-linear IC is currently available. Gaussian approximations, originally developed for linear cancellation, have been applied in an attempt to analyze performance, e.g. [234]. The results obtained for non-linear cancellation are, in general, not very accurate for more than 2-3 stages of cancellation. Even for 2-3 stages, the approximation has to be used with care.

Recently interference cancellation with the CSD function was recognized as a special cases of an iterative algorithm for solving a so-called box-constrained ML problem. Here the ML optimization is done over all vectors contained within a hypercube described by the actual data vectors. Tan et al. [265] have shown that the output of a CSD based cancellation structure is in fact the ML solution to such a problem. They have also devised necessary conditions for convergence which reveal that a CSD based SIC structure always converges in an AWGN channel. This is in contrast to a PIC structure which requires partial cancellation, or weighting similar to the over-relaxation described for the linear case for guaranteed convergence. Such weighting was first demonstrated in [56] and later in [38, 234]. In [234] some simulation studies for a radio environment have shown that weighting parameters of around 0.7 work well for a serial approach while a factor of 0.5 is required for a parallel approach.

7.5 JOINT DECODING FOR CODED CDMA

In cellular systems forward error correction coding (FEC) is used to improve the BER performance for each user. There are two possible approaches to the use of FEC and multiuser detection in a receiver:

- The multiuser detection algorithm is applied first, and the soft/hard outputs are processed by the FEC decoding algorithm.

- The FEC decoding is performed if possible within the multiuser detection algorithm, so that the data estimates that are used to estimate the MAI components are those which have been error-corrected and are thus more reliable.

In [87], the optimal ML receiver for joint decoding is proposed. As both the CDMA channel and the FEC encoders are described by finite state machines, they each impose a trellis structure on the transmission. The joint effect can be described by a joint finite state machine (FSM) with a corresponding super trellis. This joint FSM now has a memory length determined by the sum of the memory lengths of the CDMA channel and the FEC encoder, respectively. The total memory length is therefore $(K-1) + KK_{cc}$, where each individual FEC encoder has memory length K_{cc}. The number of states in the super trellis is then $2^{(K-1)+KK_{cc}}$ which, not surprisingly, leads to a computational complexity even more prohibitive than the MLSD CDMA detector.

A linear approach is taken in [8] to accommodate joint detection and decoding based on the decorrelator. This decoder is based on incorporating the linear decorrelating process into the FEC decoder by modifying the metric. In effect, the projection receiver in [8] is a decorrelating detector followed by an FEC detector based on the Mahalanobis distance [236] rather than the Euclidean distance normally used. In [133] FEC decoding is incorporated into an IC structure. By embedding Viterbi decoding within the cancellation structure, significant improvements are achieved at the expense of a substantially increased detection delay. Even further improvements are achieved by letting the bandwidth expansion be done entirely through low-rate error control coding. This is suggested in [82, 79] where a family of low-rate codes was also found through exhaustive code search based on certain appropriate distance criteria. In this case, single user performance is obtained for even highly loaded systems with only a few iterations.

The same level of performance has been reached by interpreting the bank of K single user convolutional encoders at the transmission side together with the time-varying convolutional encoder representing the CDMA channel as a serially concatenated turbo coding system. This system can be iteratively decoded based on the turbo decoding principle illustrated in Figure 7.4 [220].

However, in its pure form, the CDMA MAP decoder is prohibitively complex for practical systems. Suggestions for suboptimizing the CDMA MAP are presented in [220]. The performance is close to the single user bound, but complexity is still a problem. Since the outputs of the CDMA encoder are predominantly determined by the corresponding input symbol, it is possible, however, to suboptimize the corresponding MAP decoder into a parallel interference canceler based on probabilistic information [7]. Only marginal performance degradation as compared to optimal MAP is observed at a substantially lower complexity. This is due to the inherent power of the iterative exchange of reliability information rather than the optimality of each step. Performance approaching the single user bound is observed even for overloaded systems. Indeed, such a receiver is seen to provide levels of service within a dB of the information theoretic capacity of the CDMA channel.

An analytical approach to estimating the performance of the scheme has been suggested in [7]. It is based on Gaussian approximations and in fact, the iterative procedure works by progressively decreasing the variance of the

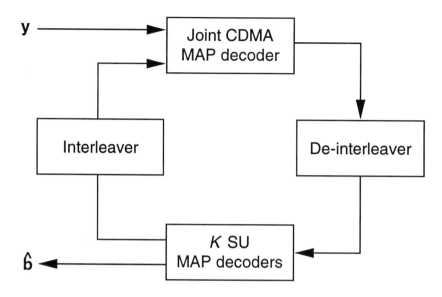

Figure 7.4. The fundamental structure of a powerful iterative receiver for joint detection and decoding.

outputs from the CDMA detector and the FEC decoders, respectively. For more detail, see [7].

7.6 NUMERICAL EXAMPLES FOR INTERFERENCE CANCELLATION

Interference cancellation has been subject to much attention for practical implementation. We therefore focus on illustrating the performance potential of such multiuser detection strategies as compared to conventional matched filter detection and single user performance. In all the simulations, the UMTS-like transmission formats, as discussed in Appendix G, are used unless otherwise noted. All simulations were performed with the WCDMA Simulation Environment software included with this book (see Appendix H). First the case of a relatively lightly loaded system over an AWGN channel are investigated. Here we have $K = 10$ users at a processing gain of $N = 32$, no FEC coding is included and only one receiving antenna element is used. The performance of different strategies is contrasted to a more highly loaded system where $K = 24$.

For the lightly loaded case, the performance of successive cancellation with various tentative decision functions is also studied over the UMTS channel model for a vehicular environment with a transmitter speed of 20 kmp/h. Finally the effects of multiple receiving antenna elements in conjunction with cancellation is investigated for the cases of $M_D = 4$ and $M_D = 8$ elements in the diversity array.

7.6.1 Interference Cancellation over an AWGN channel

Here we consider an AWGN channel without random phase rotation and fading. There are $K = 10$ users transmitting simultaneously in an asynchronous manner with a processing gain of $N = 32$. No FEC coding is included and perfect power control is assumed such that all users are received at the same amplitude level.

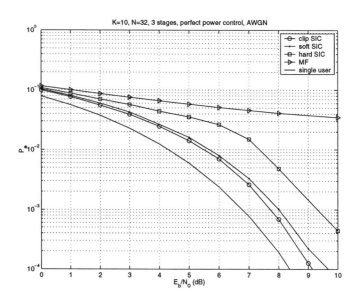

Figure 7.5. The performance of the 3 SIC schemes over an AWGN channel, $K = 10$, $N = 32$. Each SIC has 3 stages. The matched filter detector and the single user bound are included for reference.

In Figure 7.5 the performance of SIC is shown for clipped soft, soft and hard decisions, respectively. In Figure 7.6 the performance of the same detection strategies is depicted for $K = 24$.

The simulations are based on 3 stages of cancellation for each user. For comparison, the performance of the conventional detector and the single user bound are included. As we can see, significant performance improvements are provided by each cancellation strategy as compared to conventional detection. The performance of the clip SIC is however still 1 dB from single user performance. As expected, the CSD SIC provides the best performance while the hard decision case is around 2 dB worse at a BER of 10^{-3}. The soft decision case performs close to the CSD case. It is worth noting that neither of the SIC cases appear to experience an error floor above 10^{-4}.

Here the improvement achieved by using the CSD SIC is obvious. At a BER of 10^{-3}, gains in excess of 8 dB are found as compared to the other strategies. It is also noted that for an E_b/N_0 of 12-13 dB, the performance of hard SIC becomes better than soft SIC. This occurs when the noise enhancement

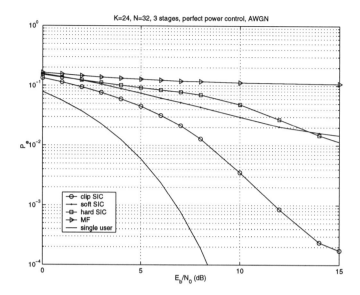

Figure 7.6. The performance of the 3 SIC schemes over an AWGN channel, $K = 24$, $N = 32$. Each SIC has 3 stages. The matched filter detector and the single user bound are included for reference.

experienced for soft (linear) cancellation becomes larger than the additional MAI created by erroneous cancellation in hard SIC. However, both soft and hard decision SIC fail to provide acceptable performance at reasonable levels of E_b/N_0. The CSD is therefore an appropriate choice as a tentative decision function for the SIC scheme. In this case it appears as an error floor around a BER of 10^{-4} is encountered for the CSD SIC. However, the loss as compared to single user performance is around 5 dB at a BER of 10^{-3}.

In Figure 7.7 the performance of corresponding PIC schemes is shown for $K = 10$. Here the CSD and hard decision PIC schemes perform virtually identically up to an $E_b/N_0 = 8$ dB. The performance is around 0.5 dB from the single user case and the soft decision PIC has a loss of around 1 dB at a BER of 10^{-3} as compared to the other two cancellation strategies. In this case, it seems as if the soft decision case suffers an error floor at around 10^{-4}.

In Figure 7.8 the case of $K = 24$ is shown.

In this case the soft PIC fails to provide acceptable performance while both hard and CSD PIC can offer reasonable BER only at $E_b/N_0 > 10$ dB. Again the CSD has the better performance of the strategies investigated. The performance is, however, significantly worse than the clip SIC. At a BER of 10^{-3}, the clip PIC has a loss of around 5 dB as compared to the clip SIC.

Comparing the performance of SIC and PIC, it is found for this UMTS based simulation that, for low SNR, PIC schemes offer slightly better performance while SIC has a steeper BER descent for moderate to high SNR. For the two

178   SPACE-TIME PROCESSING FOR CDMA

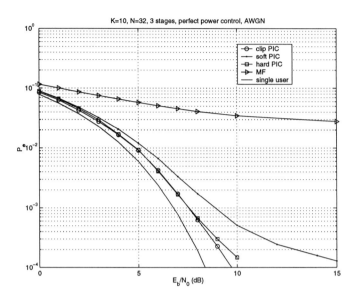

Figure 7.7. The performance of the 3 PIC schemes over an AWGN channel, $K = 10$, $N = 32$. Each SIC has 3 stages. The matched filter detector and the single user bound are included for reference.

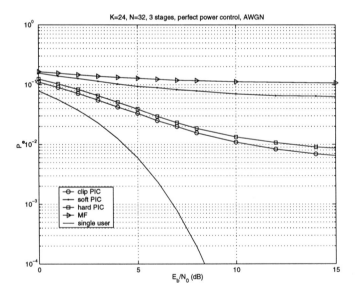

Figure 7.8. The performance of the 3 PIC schemes over an AWGN channel, $K = 24$, $N = 32$. Each SIC has 3 stages. The matched filter detector and the single user bound are included for reference.

cases presented here, the cross-over point is around 7-8 dB. This is illustrated for the case of $K = 10$ in Figure 7.9.

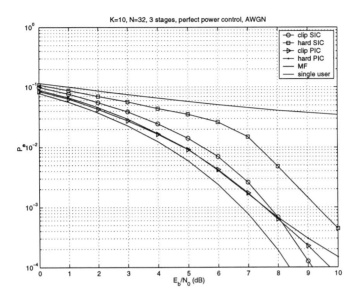

Figure 7.9. Performance comparison of PIC and SIC schemes over an AWGN channel, $K = 10$, $N = 32$. The matched filter detector and the single user bound are included for reference.

As the load increases, the CSD SIC provides increasingly better performance when compared to similar PIC schemes.

7.6.2 Interference Cancellation over UMTS Multi-path Channel Model

In this case the channel model is the UMTS model for a vehicular environment. The speed of the transmitter is 20 kmp/h and a 3-finger RAKE receiver front-end is used. Otherwise, the parameters are as specified previously. In Figure 7.10 the performance of the SIC schemes is presented.

It is now obvious that all schemes reach an error floor at around 2-5 % BER which is unacceptable. It is even questionable whether FEC coding is able to clear up raw BER at this level and provide a workable performance level.

The use of an antenna array in conjunction with interference cancellation can overcome these difficulties. In Figure 7.11 the performance of the conventional matched filter detector and the CSD SIC is shown for 1, 4 and 8 element diversity arrays.

The significant gain provided by a 4 element array as compared to the single element case is obvious. In fact, the 4 element CSD SIC provides a BER which is acceptable for uncoded performance. For an 8 element array, the CSD SIC has a performance below 10^{-4} for $E_b/N_0 > 13$ dB and even the conventional detector provides an acceptable BER for further FEC refinements.

180   SPACE-TIME PROCESSING FOR CDMA

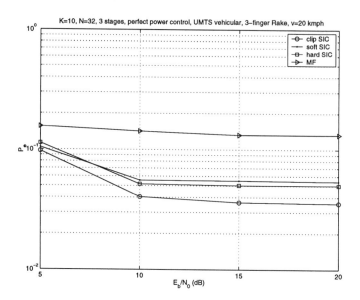

Figure 7.10. The performance of the 3 SIC schemes over the UMTS vehicular channel model, $K = 10$, $N = 32$. Each SIC has 3 stages. The matched filter detector is included for reference.

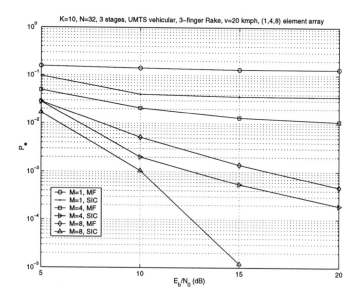

Figure 7.11. Performance comparison of the MF detector and the clip SIC over the UMTS vehicular channel model, $K = 10$, $N = 32$ with (1,4,8) element antenna arrays. Each SIC has 3 stages.

These results show that interference cancellation in conjunction with the use of diversity antenna arrays provides a powerful combination for which acceptable performance can be achieved at reasonable E_b/N_0 in an efficient way with regard to complexity.

7.7 SUMMARY

In this chapter we have provided a brief insight into the concepts and principles of multiuser detection. In our discussion here, the use of antenna arrays in conjunction with multiuser detection has been considered. As the objective of multiuser detection is to resolve MAI, a potential antenna array is most effectively used to provide maximum output SNR rather than to directly limit interference through null steering. Assuming a perfect multiuser detector that completely resolves MAI, optimal performance is only achievable if maximum SNR is guaranteed. The antenna beam should, therefore, be pointed directly at the angular location of the desired user, regardless of the interferer locations.

In our review, we have classified detector structures as either one-shot or iterative schemes. Examples of one-shot detectors are the optimal ML detector, the linear decorrelator and the linear MMSE detector where the MMSE detector can be implemented as an adaptive filter which makes it interesting for applications with short codes. All these one-shot detectors are, with the possible exception of the adaptive LMS-MMSE detector, too complex for implementation with current technology.

The class of iterative structures includes all interference cancellation schemes, both linear and non-linear. It has been noted that linear IC schemes are iterative implementations of linear, one-shot detectors, while no such connection has yet been established for non-linear schemes. Non-linear schemes, however, perform significantly better than similar linear schemes at no extra complexity costs. The CSD decision function especially has been shown to provide very good performance. The concept of weighted cancellation has also been introduced as a powerful tool for performance enhancement.

Considering the processing requirements for the different classes of multiuser detection, it is concluded that IC schemes are the most suitable structures for practical implementation. This is in agreement with the direction of most leading research groups, working on practical implementation as demonstrated in published material. Selected simulation results show that interference cancellation in conjunction with the use of antenna arrays provides a powerful combination for which acceptable performance can be achieved at reasonable E_b/N_0 in an efficient way with regard to complexity.

Notes

1. For a more general algebraic model for CDMA transmission with multiple antennas where both chip-level and bit-level signals are available for detector design, see [216].
2. The concepts of beamforming and receive- and transmit diversity are described in Chapter 5.

3. The number of paths used for combining, or equivalently multi-path fingers in the receiver, is a design parameter. Usually, either $L_R = 2$ or $L_R = 3$ fingers are considered.

4. RAKE reception is only truly MRC if there is no correlation between multi-path components of the same user. That is rarely the case so RAKE reception is not generally equal to MRC.

5. The capacity of CDMA, expressed as bits per channel use per user goes to zero as the number of users goes to infinity [46, pg. 406].

6. For certain special cases of correlation matrices, it is possible to devise a polynomial time algorithm, e.g., [233].

7. Short codes repeat themselves every T_s seconds, i.e., the same spreading code is used for every symbol interval. This is opposed to long codes where a spreading waveform with a period far greater than T_s is used. The spreading codes are then different for each symbol interval.

8. The information theoretic capacity of the CDMA channel places fundamental constraints on the number of users that can be perfectly canceled [93].

9. It should be noted that the partitioning of **A** into blocks can be arbitrary as the algorithm will still work. Different groupings do, however, affect detection delay and processing requirements. See [266] for details.

8 SPACE-TIME CODED TRANSMIT DIVERSITY FOR CDMA

In previous chapters we have seen how space-time processing can improve uplink performance and capacity of a cellular CDMA communication systems. In particular, techniques based on multi-antenna receive diversity and beamforming techniques were considered with the inclusion of multiuser detection to improve the uplink performance. Techniques to improve the downlink performance have not been developed with the same intensity to date, but is of increasing importance due to the fact that the capacity demand imposed by the projected data services, for instance internet, burdens (more heavily) the downlink channel. It is therefore of importance to find techniques that improve the downlink capacity.

Transmit diversity (see also Section 5.2.1) is an effective method to combat slow fading when multiple receive antennas are not available. Techniques such as diversity, antenna-selection, frequency-offset, phase sweeping, and delay diversity have been studied extensively in the past [102, 107, 137, 242]. Recently, space-time coding was proposed as an alternative solution for high rate data transmission in wireless communication systems [268, 243, 269, 180]. It is our objective in this chapter to consider coded transmit diversity techniques as a means to increase the downlink capacity.

In theory, the most effective technique to mitigate multi-path in a wireless channel is transmitter power control. If the channel conditions experienced by the receiver are known at the transmitter, the transmitter can pre-distort the signal in order to compensate for the distortion introduced by the channel at the receiver. Other effective transmit diversity techniques are time, frequency and space diversity.

Recent studies have explored the limit of multiple antenna system performance in a Rayleigh fading environment from an information-theoretic point-of-view [271, 76, 75]. It has been shown that, with perfect receiver channel state (side) information (CSI), and independent fading between pairs of transmit-receive antennas, the situation of total capacity may be achieved. It is clear that multiple antennas at both the transmitter and receiver, diversity transmission/reception, is one of the most effective methods to mitigate the effects of multi-path fading and, when possible, wireless communication systems should be designed to encompass all forms of available diversity to ensure optimum performance [202].

The condition of statistically independent (uncorrelated) fading is seldom achieved in practice due to the scattering environment around the mobile and basestation (as explained in Chapter 3). However, decomposition of the multiple access channel into a number of nearly independent sub-channels can be realized, provided that CSI is available at the receiver. Under these conditions (independent sub-channels), maximizing the free distance of the space-time coded symbols (transmitted over different diversity channels), a coding-diversity gain can be achieved, and is referred to as *space-time gain*. Thus, when an appropriately designed space-time code is employed, such that the space-time gain is maximized, reliable transmission of the information is guaranteed.

Foschini [75] has considered a particular layered space-time architecture with the potential to achieve higher capacity. This layered architecture forms the basis for the class of orthogonality decomposable coded space-time decoders. We will focus on the construction and performance evaluation of space-time coded CDMA employing multiple transmit antennas. The classification of space-time coded transmit diversity structures is illustrated in Figure 8.1. The techniques for transmit diversity suitable for CDMA can be divided into two distinct classes, CDTD and TDTD (see Sections 5.2.1.1 and 5.2.1.2). We discuss these two coded space-time transmit diversity classes by looking at

- the suitability of classical convolutional- and turbo codes. (The application of low rate codes, including orthogonal- and super-orthogonal convolutional codes (SOCC), is also investigated); and

- extensions of the CDTD scheme to space-time turbo diversity codes, viz. turbo transmit diversity (TTD), including both classical and super-orthogonal turbo codes (SOTC).

8.1 SPACE-TIME CODED MULTIPLE ACCESS

It is well known that CDMA systems exhibit maximum capacity potential when combined with FEC coding [304, 305, 306, 307, 308]. In fact, most FEC systems, especially those with low code rates, expand bandwidth and can be viewed as spreading systems. For a fair comparison (in terms of equal information

SPACE-TIME CODED TRANSMIT DIVERSITY FOR CDMA 185

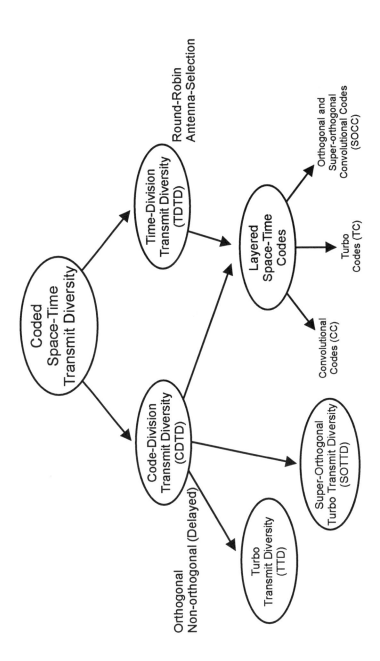

Figure 8.1. Coded transmit diversity space.

throughput) of low code rate systems to uncoded systems, the spreading sequence length N, must be shortened by a factor of R_c, the code rate. The positive trade-off between greater distance properties of lower rate codes and increased cross-correlation effects (due to shorter sequence length) is fundamental to the success of coded CDMA. We also know from information theory that the maximum theoretical CDMA capacity can only be achieved by employing very low rate FEC codes utilizing the entire bandwidth, without further spreading by the multiple access sequence [112, 307, 80, 81].

Viterbi [304, 307] has proposed the use of orthogonal convolutional codes as low rate code extensions for code-spreading CDMA. Recently, two new classes of low rate codes with improved performance have been proposed. Pehkonen *et al.* [196, 197] proposed a coding scheme that combines SOTCs with SOCCs [307]. A different approach was taken by Frenger *at al.* [80, 81], where a class of nested rate-compatible convolutional codes (RCCC), with maximum free distance (MFD), was derived.

In [34], performance gains achieved in a RAKE based CDMA system with convolutional versus trellis coding were reported. Codes were constructed over an MPSK signal set by taking a standard Ungerboeck type code for MPSK modulation and multiplied by a binary pseudo-noise sequence, thereby spreading the signal over a large bandwidth. It was reported that this approach did not yield a performance advantage over standard convolutional codes, with the conclusion that it is better to exploit the low distance properties of low rate convolutional codes as oppose to using higher order modulation schemes for efficient signalling.

A different approach to trellis coded CDMA was investigated by Woerner *et al.* [324]. In this approach the trellis code is constructed over the set of possible signature sequences rather than over some $2D$ signal constellation. Instead of expanding the number of signal points in the $2D$ constellation, the signal points were expanded over a set of orthogonal spreading sequences. A carefully designed trellis then allows only certain combinations of sequences that have a large total minimum distance. By increasing the number of sequences, the actual minimum distances between sequences have been decreased. The trellis code compensates for this decrease by increasing the minimum distance of the code above that of the uncoded system.

For non-optimum multiuser receivers, such as the MF or RAKE, coding gain comes at the cost of increased MAI level. A limitation to the use of low rate coding comes when the spreading is reduced to such a level that the MAI does not appear Gaussian anymore. When transmit diversity is considered, this situation (from a coding perspective) is improved due to the additional MAI. Especially, when turbo coding is considered, the potential coding gain can be substantial. By using more powerful codes than those used by Boudreau *et al.* [34], the issue of spreading versus coding can be more adequately addressed. For a finite effective code rate (and hence a finite spreading ratio), the level of MAI, under AWGN, equal power conditions, is fixed. If the MAI was truly Gaussian in nature, turbo codes should perform in a similar way as if applied to an AWGN channel. For a RAKE receiver with perfect channel estimation, the soft input turbo code will perform equally well in an AWGN and a fading channel. The power of turbo coding approaching the Shannon bound in narrow band systems, implies that almost optimum performance should be achievable with coded CDMA systems under similar signalling conditions.

In the next two section background on space-time coding for CDMA is presented, together with a review of FEC codes. Specific emphasis on turbo coding concepts required for the design and analysis of space-time coded systems are presented. This is followed by a comprehensive discussion of layered space-time coding for CDMA.

8.2 TURBO CODING

FEC coding is an extremely complex topic to which entire books are dedicated. We will confine our discussion to encoding and decoding configurations, ranging from classical to more advanced turbo code extensions, applied to coded space-time transmission systems. To provide a basis for understanding turbo codes, we will review some of the fundamental concepts. The reader is referred to Lin and Costello [154] or Clark and Cain [44] for detailed introductory discussions on FEC coding, and to Viterbi and Omura [309], Petersen and Weldon [198], Gallager [86], and Berlekamp [24, 26] for advanced discussions of both information theoretic and algebraic foundations of FEC coding.

In their quest to design good FEC codes, coding theorists have developed codes with a high amount of structure, which lends itself to feasible decoder implementation. Although coding theory suggests that codes chosen "at random" should perform well if the block size is large enough, it is mainly the high complexity of conventional ML decoding that poses the real challenge in decoding these "almost" random codes.

Since conventional block- and convolutional codes are highly structured, encoders and decoders with reasonable implementation complexity is possible. However, the very same structure that facilitates practical implementation, results in significantly inferior performance gains relative to the random coding bounds predicted by Shannon.

With this in mind, perhaps the most exiting and potentially important development in coding theory in recent years has been the introduction of parallel concatenated convolutional codes by Berrou *et al.* [29]. The term "turbo code" was adopted to describe this new class of code. The introduction of turbo coding has opened a whole new way of looking at the problem of constructing good codes with low complexity decoding. While turbo codes still contain enough structure to admit practical encoding and decoding algorithms, turbo codes possess *random-like* properties. As a consequence, the performance of turbo codes comes much closer to the Shannon bound than conventional block and convolutional codes.

8.2.1 Turbo Encoding

The turbo encoder is composed of two or more recursive systematic convolutional (RSC) encoders, which are in general identical. The constituent encoders receive the "same" data, the only difference is that the stream to each encoder is permuted by an interleaver, with the result that turbo codes appear random. Because the interleaver must have a fixed structure and generally works on data in a block-wise manner, turbo codes are by necessity block codes.

Recall that the minimum distance of a linear block code is a good first order estimate of the code's performance. For linear block codes, the minimum distance is the smallest non-zero Hamming weight of all the valid code words. The combination of interleaving and RSC encoding ensures that most code words produced by a turbo code have a high Hamming weight. Because of its infinite impulse response properties, the output of an RSC encoder generally has a high Hamming weight. There are, however, some input sequences which cause an RSC encoder to produce low weight outputs. Because of the interleaver, the two RSC encoders do not receive their inputs in the same order. Thus, if one encoder receives an input that causes a low weight output, then it is improbable that the other encoder also receives an input that produces a low weight output. Unfortunately, there will always be a few input messages that cause both RSC encoders to produce low weight outputs and thus the minimum distance of a turbo code is not, in general, particularly high. But the *multiplicity* of low weight code words in well designed turbo codes is low. It is because of the small number of low weight code words that turbo codes can perform well at low SNR [282].

However, the performance of turbo codes at higher SNRs becomes limited by the relatively small minimum distance of the code. While the goal of traditional code design is to increase the minimum distance of the code, the goal of turbo code design is to reduce the multiplicity of low weight code words.

8.2.2 Turbo Decoding

The problem of estimating the states of a Markov process observed through noise has two well known trellis-based solutions — the Viterbi algorithm [74] and the (symbol-by-symbol) maximum *a posteriori* (MAP) algorithm [14, 15]. The two algorithms differ in their optimality criterion. The Viterbi algorithm finds the most probable transmitted sequence, while the MAP algorithm, on the other hand, attempts to find the most likely transmitted symbol, given the received sequence [282].

The soft output Viterbi algorithm (SOVA) is an extension of the classic Viterbi algorithm that provides the reliability of the bit estimates. In addition, the improved SOVA algorithm, utilizing a multiplicative correction factor to improve the reliability estimates, may also be considered.

An excellent overview of the trellis-based soft-input soft-output (SISO) decoding algorithms has been presented in [282]. A discussion on the MAP, Max-Log-MAP and Log-MAP are presented here. The MAP algorithm calculates the *a posteriori* probabilities directly. However, the algorithm suffers from a high computational complexity and numerical sensitivity. The max-log-MAP and log-MAP algorithms perform the MAP algorithm in the log domain, which significantly reduces complexity and numerical sensitivity.

In Appendix F the the underlying concepts and implementation issues of iterative "turbo" decoding by means of the algorithm is explained in detail.

8.2.3 Turbo Code Performance

Figure 8.2 depicts the turbo code design space [28]. The design space can be grouped into a service dependent and an implementation dependent components. The service dependent components influence typically the quality of service and the data rate. The implementation dependent components influence the maximum decoding delay, the implementation complexity, system flexibility, modularity and integratability.

Below, a short description of the more important blocks of Figure 8.2 is given.

8.2.3.1 Turbo Interleaver/Permuter.
The interleaver (or permuter) component of the turbo encoder directly defines the service dependent part of the system design space. The weight distribution of the codewords produced by the turbo decoder depends on how the codewords from one of the simple components are teamed with codewords from the other encoder(s). Stated differently, the performance of the turbo code depends on how effectively the data sequences that produce low encoded weights at the output of one encoder, are matched with permutations of the same data sequence that yield higher encoded weights at the outputs of the others. Two characteristics of the interleaver is of particular importance

- Interleaver size, N_{tc}. This is the most important factor influencing the turbo code performance, and it is well known that performance improves as the interleaver size increases [21]. The gain, in terms of error performance, with increased interleaver is formally known as the *interleaver gain*. As interleaver size (gain), however, increases so does decoding delay, and a balance must be found between acceptable performance and tolerable latency. At high SNRs, the interleaver design becomes critical [22], where the performance is dominated by the low weight code words. At low SNRs, turbo codes perform well with almost any (randomly permutated) interleaver, provided that the inputs at the RSC encoders are sufficiently uncorrelated.

- Interleaver selection. If randomly chosen permutations perform well, then in principle it is possible to design deterministic permutations that work even better. In [55], several non-random permutations have been investigated. These are: (i) permutations based on block interleavers, (ii) permutations based on circular shifting, and (iii) semi-random ("S-random") permutations.

190 SPACE-TIME PROCESSING FOR CDMA

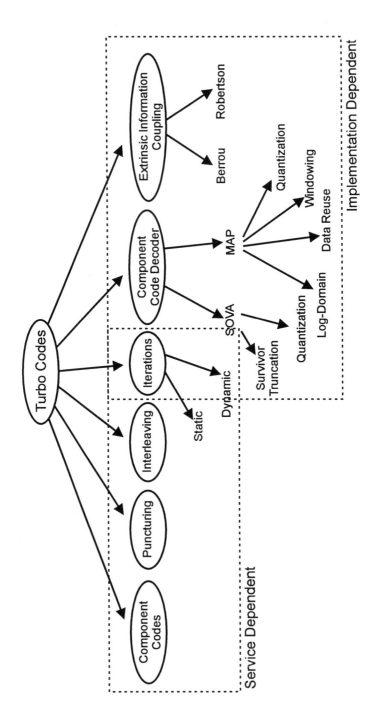

Figure 8.2. Turbo coded system design space [28].

8.2.3.2 Constituent Encoder.
As discussed previously, encoders being recursive and systematic are normally chosen as constituent encoders. In [55] it was argued that recursive encoders do not modify the output weight distributions of the individual component codes, but only change the mapping between the input data and output encoded sequences. In addition, for a non-recursive encoder, nearly all low weight input sequences are self-terminating. As a result, the output weight is strongly correlated with the input weight for all possible input sequences. It is precisely this characteristic that makes these encoders very undesirable as constituent component encoder. The motivation for systematic encoders stems from the fact that puncturing can be employed to realize code rates higher than achievable with non-systematic encoders.

Surprisingly, the choice of constituent RSC encoder, and in particular its constraint length, does not significantly influence the performance of turbo codes. For this reason, turbo codes typically use simple constituent codes with constraint lengths of $K_{tc} = 3, 4$, or 5.

8.2.3.3 Puncturing.
As for most other codes, performance degrades as code rate increases. If puncturing is used to increase the code rate, then the manner of puncturing is also a performance factor and puncturing matrices may need to be considered. The joint design of interleavers and puncturing matrices is perhaps the most important aspect of turbo code design. The puncturing system also directly define the service dependent part of the system design space.

8.2.3.4 Decoding Algorithm.
Most decoding algorithms are iterative, and therefore the number of iterations has an impact on the performance. The number of iterations is static or determined dynamically during decoding after evaluation of some criteria [224, 225]. Decoding is normally performed with the MAP or SOVA algorithms. When implementing the SOVA algorithm, the designer has to choose among several implementation options to reduce computational complexity, increase throughput, or reduce power consumption.

Extrinsic information coupling (see also Appendix F) is typically performed according to Berrou's original method [29] or directly, which has first been proposed by Robertson [225].

8.3 SPACE-TIME CODED SYSTEM MODEL

Figure 8.3 illustrates the components of a space-time coded CDMA system. A discrete memoryless source (DMC) is designed to output a sequence of independent, equally likely symbols $\mathbf{b} = \ldots, b_1, b_2, b_3, \ldots$ from the alphabet $\{0, 1\}$. The space-time channel coder receives a sequence of input symbols $\mathbf{b}$ from the DMC and outputs a sequence of symbols $\mathbf{x} = \ldots, x_1, x_2, x_3, \ldots$ from the alphabet $\{0, 1, \ldots, M_c\}$, where M_c is the number of symbols available for transmission.

The sequence of space-time encoder output symbols has a very carefully controlled structure which enables the detection and correction of transmission

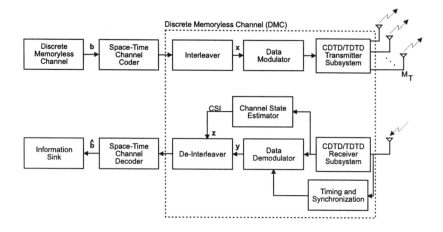

Figure 8.3. Block diagram of a space-time coded transmit diversity CDMA system.

errors in a multiple transmit antenna signalling. The coded space-time encoder should be designed in such a way that the combined spatial and temporal properties will guarantee maximum transmit diversity, while maintaining the option to include receive diversity. The channel coding may be either convolutional- or turbo coding.

After encoding, the output is split into 1 (for TDTD) to M_T (for CDTD) streams and each of the streams are independently interleaved, using a block symbol-by-symbol interleaver. The function of the interleaver/de-interleaver is to distribute channel errors randomly throughout the decoder input sequence **y**, thereby enabling the use of coding, optimal for AWGN, to function well under adverse MAI and multi-path fading.

The multiuser code/time division transmit diversity sub-system generates a signalling waveform based on the combinations of data modulation, spreading modulation, coding and transmit diversity schemes. The transmit diversity sub-system is formed by either CDTD (see Section 5.2.1.1) or TDTD (see Section 5.2.1.2).

Since the signal at the receive antenna is a linear superposition of the $K \times M_T$ transmitted orthogonal signals, the receiver first performs chip waveform matching with reference to the M_T streams associated with the desired user. This despreading operation is the key function of any spread-spectrum system, and can be accomplished only if accurate synchronization information is available. Spreading waveform synchronization, carrier recovery, symbol and frame synchronization are beyond the scope of this chapter, but standard schemes can used [205]. Channel estimation is performed on each resolved path, and used in the pilot symbol assisted (PSA) RAKE combiner to resolve each of the transmitted streams from the multiple transmit antennas.

SPACE-TIME CODED TRANSMIT DIVERSITY FOR CDMA   193

The total of M_T baseband *soft-decision* streams formed by the demodulator are then de-interleaved using a block symbol-by-symbol de-interleaver. The de-interleaved output sequence is denoted by $\mathbf{y} = y_1, y_2, y_3, \ldots$, which is input to the ML sequence space-time decoder. ML decoding is optimum in terms of achieving the lowest error probability.

To assist the space-time decoder, additional reliability information can be obtained by measuring the CSI. The output of the CSI is denoted $\mathbf{z} = z_1, z_2, z_3, \ldots$ where z_j is a scaled real value.

All the functions within the dashed box of Figure 8.3 define a DMC with input vector $\mathbf{x}$ and output vectors $\mathbf{y}$ and $\mathbf{z}$. The DMC may be completely characterized by the probability that the channel output is $\mathbf{y}$, given that the channel input is $\mathbf{x}$ and $\mathbf{z}$ is the CSI. That is, the channel is completely characterized by

$$p(\mathbf{y} \mid \mathbf{x}, \mathbf{z}) = \prod_{i=1}^{\infty} p(y_i \mid x_i, z_i), \qquad (8.1)$$

where $p(y_i \mid x_i, z_i)$ is the probability that, given the input x_i and the CSI z_i, the demodulator output is y_i for channel i.

The probabilities $p(y_i \mid x_i, z_i)$ are found by analysis of the data modulation/demodulation, and the waveform channel, including transmit diversity. When the demodulator output is continuous, the probabilities of (8.1) are replaced by continuous probability density functions. Characterization of the channel using (8.1) enables decoupling the analysis of the waveform channel from the FEC analysis.

The principle goal of the remainder of this chapter is to consider in detail the construction and performance evaluation of layered space-time codes for CDMA communication systems operating on frequency selective multiple access channels. This will be accomplished using the space-time coded system model described in this section. In the analysis perfect synchronization is assumed and it is further assumed that the MAI is Gaussian distributed (making use of the Gaussian assumption). In addition, perfect power control is assumed, implying that the base station adjusts the transmitted power such that the mobile terminal observes a prescribed SINR.

8.4 LAYERED SPACE-TIME CODES

Here we consider the construction of layered space-time convolutional- and turbo coding configurations applied to multiple transmit antenna signalling.

Figure 8.4 illustrates the general block diagrams of the layered space-time encoder and decoder. In this model the data source produces a sequence of N_{cc} data bits, $\mathbf{b}_k$, which enters a rate R_c encoder. The encoded sequence, $\mathbf{x}(j)$ of length N_{cc}, is passed to the M_T symbol interleavers and to the multiple access transmit diversity sub-system. Recalling from Figure 8.1, this coding scenario supports both the CDTD and TDTD signaling configurations.

In the receiver subsystem, the received signal is correlated with the complex scrambling/spreading sequence associated with each of the individual an-

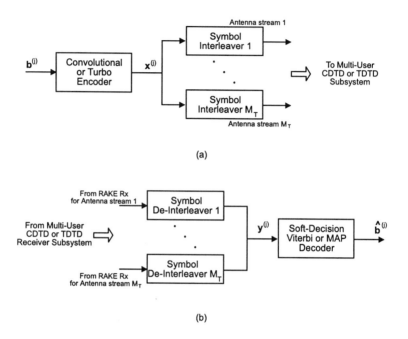

Figure 8.4. Generalized block diagram of the space-time convolutional coder based on a sub-optimum configuration, (a) Encoder, (b) Decoder.

tenna sub-streams. With the decoding configuration, as shown in Figure 8.4(b), nearly optimal decoding can be achieved by employing the Viterbi or MAP algorithm. This decoding is suboptimal since the correlator receiver is matched to the AWGN channel, and not to the MAI. It has been shown that when CSI information is available at the receiver, using soft-decision decoding and coherent detection, the coding gain on fading channels can be quite high.

It should be noted that for TDTD signalling, the configuration presented in Figure 8.4 can accomplish a significant portion of the theoretical system capacity, although it is a suboptimal implementation.

In contrast to the suboptimal TDTD configuration, in CDTD the M_T orthogonal streams are recovered from the receiver sub-system by means of a decorrelation process. During a specific decoding stage, the decorrelation detector only considers a single user entity for the spreading sequences associated with the M_T transmit antennas associated with the reference user. The CDTD scheme may be extended to a single user multiple antenna decoder based on a optimum successive decoder (OSD). Figure 8.5 illustrates a possible implementation of this decoder for convolutional codes, where each user is decoded on a per-user channel basis.

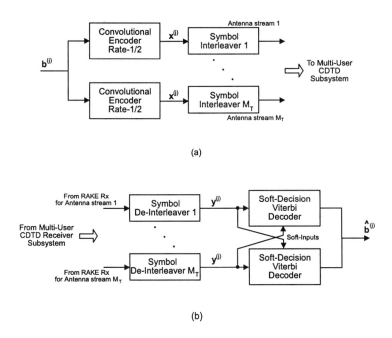

Figure 8.5. Generalized block diagram of the space-time convolutional coder based on a per-user configuration, (a) Encoder, (b) Decoder, with soft-information transfer.

In order to maintain soft failure diversity and to provide additional soft-decision information to the different decoders operating in parallel, systematic convolutional encoders are generally utilized at the encoder and decoder. In this way, the full benefits of soft-decision decoding are realized. In [293, 69], a group metric decoding scheme has been proposed that exploits the nature of the multiuser decoding problem. This decoder may be considered as a single user, single antenna decoding scheme, that will utilize information from all the user and antenna matched-filter outputs to the decoding metrics.

8.4.1 Low Rate Code Extension

The low rate codes proposed in [304, 334, 222] for CDMA, can be directly applied to configurations depicted in Figures 8.4 and 8.5. Here we consider the use of orthogonal and super-orthogonal convolutional codes (SOCC) for this purpose.

Orthogonal block codes are known to perform well on very noisy channels. In [304], Viterbi presented a method to find orthogonal convolutional codes having similar properties. However, orthogonal convolutional codes imply a large bandwidth expansion. Along these lines, several related coding schemes with good distance properties, but less bandwidth requirements, have been proposed in [334, 222].

A rate $R_c = 1/n$ orthogonal convolutional encoder with constrained length, K_{oc}, can be constructed from a shift register and a block orthogonal encoder or signal selector [304]. One in $2^{K_{oc}}$ orthogonal waveforms is chosen based on the state of the shift register. The weight of any trellis branch (not the all-zero path) is $2^{K_{oc}-1} = n/2$. For an orthogonal convolutional code the rate is then related to the constraint length by $R_c = 2^{K_{oc}}$. One way of implementing an orthogonal convolutional encoder is for the orthogonal code selector to choose rows from a Walsh-Hadamard matrix.

A rate $R_c = 2^{-(K_{oc}-2)}$ SOCC is obtained by modifying the orthogonal codes as follows [308]

- Let the inner $K_{oc} - 2$ stages of the shift register be used for the orthogonal waveform selection; then

- add the first and last bits of the register modulo-2 to each bit in the orthogonal waveform.

Section 8.8.2 will consider the application of SOTC to space-time transmit diversity.

8.5 PERFORMANCE EVALUATION

Space-time coded performance can be calculated by extending the space-time mathematical model derived in Chapter 2 and using well known error control bounding techniques. We will also consider space-time coded performance on fading channels, where we will make extensive use of the derivation presented in Appendix D for the pdf of correlated multivariate gamma distributed random variables. Performance bounds for convolutional- and turbo codes will be presented.

8.5.1 Convolutional Code (CC) Bounds

The bounds presented here are based on block error probability bounds, originally derived by Shannon [246]. Specifically, to determine upper bounds on BER with convolutional encoding and ML decoding represented by an equivalent (n, k) linear block code, it is useful to recall the state diagram and associated generating function approach. Due to code linearity, we assume that the all-zero message has been transmitted, and we can write the upper bound on the word error probability as

$$P_w \leq \sum_{d=d_{free}}^{n} A_d \, P_d(\mathbf{c} \to \hat{\mathbf{c}}), \qquad (8.2)$$

where A_d is the number of codewords with Hamming weight d, obtained from the series expanded transfer function $T(I, D)$, and given by

$$\left.\frac{\delta T(I,D)}{\delta I}\right|_{I=1} = \sum_{d=1}^{n} A_d D^d, \qquad (8.3)$$

where D is the channel parameter. By setting $I = 1$ after differentiation, the number of bit errors, corresponding to an error event of length d, is obtained as the multiplicity term of D^d, where $D = e^{-R_c E_b/N_o}$.

The *free distance*, denoted by d_{free}, of any code is the minimum Hamming distance between any two distinct code sequences. D is a function only of the channel transition probabilities and the message decoding metric.

The conditional pairwise error probability, $P_d(\mathbf{c} \to \hat{\mathbf{c}})$, is the probability of incorrectly choosing a codeword with weight d. That is, the probability in which the incorrect encoded sequence $\hat{\mathbf{c}}_n = (c_1, c_2, \ldots, c_n)$ is chosen instead of the correct encoded sequence $\mathbf{c}_n = (c_1, c_2, \ldots, c_n)$.

For the continuous output soft-decision AWGN channel it can be shown that the single user $P_d(\mathbf{c} \to \hat{\mathbf{c}})$ is given by [303, 309, 308]

$$P_d(\mathbf{c} \to \hat{\mathbf{c}}) = Q\left(\sqrt{\frac{2dR_c E_b}{N_o}}\right). \qquad (8.4)$$

From [55], the more general expression for the average weight distribution can be written as

$$A_d = \sum_{i=1}^{k} \binom{n}{i} p(d \mid i), \qquad (8.5)$$

where $\binom{k}{i}$ is the number of input words with Hamming weight i and $p(d \mid i)$ is the probability that an input word with Hamming weight i produces a codeword with Hamming weight d. Substituting (8.5) into (8.2), the upper bound on the word and bit error rate can be expressed as

$$\begin{aligned} P_w &\leq \sum_{d=d_{free}}^{n} A_d\, P_d(\mathbf{c} \to \hat{\mathbf{c}}) \\ &= \sum_{d=d_{free}}^{n} \sum_{i=1}^{k} \binom{k}{i} p(d \mid i)\, P_d(\mathbf{c} \to \hat{\mathbf{c}}) \\ &= \sum_{i=1}^{k} \binom{k}{i} E_{d\mid i}\{P_d(\mathbf{c} \to \hat{\mathbf{c}})\}, \end{aligned} \qquad (8.6)$$

and

$$P_e \leq \sum_{i=1}^{k} \frac{i}{k} \binom{k}{i} E_{d\mid i}\{P_d(\mathbf{c} \to \hat{\mathbf{c}})\}. \qquad (8.7)$$

In (8.5) and (8.7), $E_{d|i}\{\cdot\}$ is an expectation with respect to the distribution $p(d \mid i)$. This average upper bound is attractive because relatively simple schemes exist for computing $p(d \mid i)$ from the state transition matrix of the RSC [101, 55]. This information is implicit in the generating function $T(I, D)$ associated with the particular code employed.

An expression for the SNR for coded asynchronous CDMA, denoted by Γ_c, can be determined as

$$\Gamma_c = \left(\frac{1}{R_c} \frac{N_o}{2 E_b} + \frac{(K \cdot M_T - 1)}{3N} \right)^{-1}, \qquad (8.8)$$

for CDTD, and

$$\Gamma_c = \left(\frac{1}{R_c} \frac{N_o}{2 E_b} + \frac{(K - 1)}{3N} \right)^{-1}, \qquad (8.9)$$

for TDTD. With reference to (6.17), we can see that (8.8) and (8.9) are extensions of our previous analysis and includes the code rate R_c.

For a fair comparison to an uncoded system under equal throughput and bandwidth conditions, the spreading sequence length, N, of the coded system must be shortened by a factor of $1/R_c$. This results in a degradation due to the MAI since it is well-known that the cross-correlation between any two spreading sequences is proportional to $1/\sqrt{N}$. A trade-off between the greater distance properties of low rate codes and increased cross-correlation effects (due to shorter sequence lengths) is fundamental to the success of coded CDMA.

8.5.1.1 Evaluation of $P_d(\mathbf{c} \to \hat{\mathbf{c}})$. Under conditions of fast fading, it is generally assumed that the fading is independent in successive signaling intervals. As a result, the sequence of fading amplitudes β_i constitutes an identically distributed sequence[1].

Consider the situation where the all-zeros codeword $\mathbf{c} = \mathbf{0} = \mathbf{c_0}$, is being transmitted and codeword $\hat{\mathbf{c}} = \mathbf{c_n}$ is being received. In addition, we consider a trellis path which re-emerges with the correct all-zero path, having diverged at some point in the past, and differing from the all-zero path in exactly d symbol positions. Define the n-vector

$$\mathbf{S_n} = (S_{n1}, S_{n2}, \ldots, S_{nd}), \qquad (8.10)$$

where $S_{ni}, (i = 1, 2, \ldots, d)$ represents the value of the resulting envelope power process in the ith signaling interval among those where the path differs from the all-zero path. Assuming perfect phase tracking of the phase perturbation process and CSI at the receiver, the conditional pairwise error probability for an incorrect sequence with d error symbols is [101]

$$P_d(\mathbf{c} \to \hat{\mathbf{c}} \mid \mathbf{S_n}) = Q\left(\sqrt{\Gamma_c \sum_{i=1}^{d} s_{ni}}\right), \qquad (8.11)$$

where Γ_c is the effective signal-to-noise ratio, defined in (8.8) for CDTD and (8.9) for TDTD, respectively. The average error event probability can then be determined by averaging over the random n-vector $\mathbf{S_n}$ with the result

$$P_d(\mathbf{c} \to \hat{\mathbf{c}}) = E_{\mathbf{S_n}}\left\{Q\left(\sqrt{\Gamma_c \sum_{i=1}^{d} s_{ni}}\right)\right\}, \qquad (8.12)$$

where the expectation operator $E_{\mathbf{S_n}}\{\cdot\}$ represents joint expectation with respect to the received signal power components. If it is assumed that the M_T transmissions are equal powered, with constant correlation between the branches, and transmitted over a Rayleigh fading channel, the components of the received power vector $\mathbf{S_n} = \beta_n{}^2$ are identically distributed, with pdf given by (see Appendix D equation (D.18))

$$p_{\mathbf{S_n}}(\mathbf{s_n}) = \frac{1}{\Omega^2 \Gamma(M_T \cdot L_R)} \left(\frac{\mathbf{s_n}}{\Omega^2}\right)^{M_T \cdot L_R - 1}$$
$$\times \frac{\exp\left(-\frac{\mathbf{s_n}}{(1-\rho)\Omega^2}\right) \cdot {}_1F_1\left(1, M_T \cdot L_R, \frac{\rho M_T \cdot L_R \mathbf{s_n}}{(1-\rho)(1-\rho+\rho M_T \cdot L_R)\Omega^2}\right)}{(1-\rho)^{(M_T \cdot L_R - 1)}(1-\rho+\rho M_T \cdot L_R)}, \qquad (8.13)$$

where ${}_1F_1(\cdot)$ is the confluent hyper geometric function [2], and as before, Ω^2 is the average received path strength (assumed equal, i.e. $\Omega_n^2 = \Omega^2$) and ρ the correlation between transmit or receive branches.

For fast fading with perfect CSI, the pairwise error probability is given by

$$P_d(\mathbf{c} \to \hat{\mathbf{c}} \mid \mathbf{S_n}) = Q\left(\sqrt{\Gamma_c \sum_{i=1}^{d} s_{ni}}\right). \qquad (8.14)$$

By averaging the pairwise error probability, (8.14), over (8.13), we have a multi-dimensional integral given by

$$P_d(\mathbf{c} \to \hat{\mathbf{c}}) = \int_{s_1} \int_{s_2} \cdots \int_{s_d} Q\left(\sqrt{\Gamma_c \sum_{i=1}^{d} s_{ni}}\right) \qquad (8.15)$$
$$\times \; p_{S_{n1}}(s_{n1}) \, p_{S_{n2}}(s_{n2}) \, \cdots \, p_{S_{nd}}(s_{nd}) \, ds_{n1} \, ds_{n2} \, \cdots \, ds_{nd}.$$

If the fading powers are independent, the indexes of the differing bit positions are of no importance, since only the incorrect codeword weight matters [101].

The exact evaluation of (8.15) is very difficult. To solve this problem, Hall et al., examined four options [101]. The first option is to simplify (8.15) to a form that can be evaluated via numerical integration [49]. The other three options examined, avoids the problem of numerical integration by seeking closed form upper bounds for $P_d(\mathbf{c} \rightarrow \hat{\mathbf{c}})$.

The first option proposed by Hall et al. has been used for analytical results. From [49], $Q(x)$ can be expressed in the alternative form written as

$$Q(x) = \frac{1}{\pi} \int_0^{\pi/2} e^{-x^2/(2\sin^2 \psi_h)} \, d\psi_h. \tag{8.16}$$

Substituting (8.16) into (8.14), we can write

$$P_d(\mathbf{c} \rightarrow \hat{\mathbf{c}} \mid \mathbf{S_n}) = \frac{1}{\pi} \int_0^{\pi/2} \exp\left(-\frac{\Gamma_c \sum_{i=1}^d s_{ni}}{2 \sin^2 \phi}\right) d\phi. \tag{8.17}$$

Since all the fading powers are independent, the d-dimensional integral of (8.15) reduces to a product of integrals over each S_{ni}.

Slow fading occurs when the symbol signalling rate is greater than the fading rate. That is, when $S_{ni} = \beta_{ni}^2 = \beta^2 = S$, i.e. the effective fading amplitude is assumed to be constant throughout the message sequence. The power of the received signal power S is again given by (8.13), where, for slow fading, $\mathbf{S_n} = S$.

It follows from (8.11) that

$$P_d(\mathbf{c} \rightarrow \hat{\mathbf{c}} \mid S) = Q\left(\sqrt{d\, \Gamma_c\, s}\right). \tag{8.18}$$

Making use of the inequality

$$Q(x) \le \frac{1}{2} e^{-x^2/2}, \quad x \gg 1, \tag{8.19}$$

the upper bound for the pairwise error probability can be written as

$$P_d(\mathbf{c} \rightarrow \hat{\mathbf{c}}) \le E_\mathbf{S} \left\{ \frac{1}{2} \exp\left(-\frac{1}{2} d\, \Gamma_c s\right) \right\}. \tag{8.20}$$

Using (8.7), the pairwise error probability, averaged over the fading statistics can be written as

$$P_d(\mathbf{c} \rightarrow \hat{\mathbf{c}}) \le \frac{1}{2} \int_s \exp\left(-\frac{1}{2} d\, \Gamma_c s\right) p_S(s) \, ds. \tag{8.21}$$

Using (8.7) and (8.17) for fast fading and (8.21) for slow fading, it is now possible to calculate the performance of a space-time convolutionally coded system.

8.5.2 Turbo Code (TC) Bounds

For a turbo code with a fixed interleaver, the construction of A_d (from (8.5)) requires an exhaustive search. The latter lead to the proposition of an average upper bound constructed by averaging over all possible interleavers [55]. Therefore, to derive this average performance an abstract interleaver, called the *uniform interleaver*, is used, i.e. an interleaver that, for a given input block of n bits with input weight i, outputs all $\binom{n}{i}$ distinct permutations with equal probability.

The transfer function $T(I, D)$ has been determined by Benedetto et al. [23] for turbo codes in three co-decoding configurations: namely, the transfer function of the hyper-trellis has been evaluated in conjunction with continuous, trellis truncated and trellis terminated co-decoding. The latter showed that the performance of the truncated encoder is significantly worse than that of continuous decoding, whereas trellis termination is only slightly worse.

Using (8.7), with $p(d \mid i)$ known, the performance of space-time turbo codes can be evaluated for various channels and transceivers by formulating the conditional pairwise error probability, $P_d(\mathbf{c} \to \hat{\mathbf{c}})$, for the configuration of interest [292, 290].

The expressions for $P_d(\mathbf{c} \to \hat{\mathbf{c}})$ derived for convolutional codes, are limited to the case where the output codeword weight, d is fixed. Here, we extend the results to include the performance of turbo codes where the code weight is described in terms of an input-output conditional probability density function (CPDF), $p(d \mid i)$. In Appendix E, the unconditioned (also known as the Divsalar CDPF) and conditioned (also known as the Binomial CPDF) CPDFs have been derived from the constituent encoder state transition matrix, $t(l, i, d)$.

Assuming an AWGN channel, the pairwise bit error probability of turbo codes may be written as

$$P_d(\mathbf{c} \to \hat{\mathbf{c}}) = E_{d|i}\left\{Q\left(\sqrt{d\,\Gamma_c}\right)\right\}, \tag{8.22}$$

where the conditional expectation $E_{d|i}\{\cdot\}$ is over the evaluated CPDF, $\tilde{p}(d \mid i)$, as explained in Appendix E.

Employing similar arguments to that used in Section 8.5.1.1, the bound given by (8.22) can be extended to include the transmit diversity signalling and multi-path fading channel effects.

From (8.20), for the slow fading channel, the upper bound for the pairwise error probability can be written as

$$P_d(\mathbf{c} \to \hat{\mathbf{c}}) \leq E_\mathbf{S}\left\{E_{d|i}\left\{\frac{1}{2}\exp\left(-\frac{1}{2}d\,\Gamma_c s\right)\right\}\right\}. \tag{8.23}$$

Note that every non-zero codeword is included in the above summation and where the conditional expectations, $E_\mathbf{S}$ and $E_{d|i}\{\cdot\}$ is over the instantaneous fading power pdf, and the CPDF ($\tilde{p}(d \mid i)$), respectively.

Using (8.17), the bounds of (8.23) can be extended to include fast fading.

8.6 ANALYTICAL RESULTS

In the results that follow an approximately fixed bandwidth comparison is made between coded and uncoded system performance. This results in $N = \{32, 16, 10, 8, 6, 5\}$ for code rates of $R_c = \{1, 1/2, 1/3, 1/4, 1/5, 1/6\}$, respectively. In both AWGN and 2-path (equal power) Rayleigh fading (fast and slow), while in the fading environment a RAKE receiver with $L_R = 2$ branches is assumed. We consider only the use of a single receive antenna ($M_D = 1$).

When coding is considered, constraint lengths $K_{cc} = 9$ and $K_{tc} = 3$ are assumed for convolutional- and turbo coding respectively. The generator polynomial for the convolutional code is given by $(561)_8, (753)_8$ for the rate $R_c = 1/2$ codes, and $(557)_8, (663)_8, (711)_8$ for the rate $R_c = 1/3$ codes, respectively. For the 4-state turbo code the feedforward and feedback generator polynomials are given by $g_{ff} = 5_8$, and $g_{fb} = 7_8$, respectively.

As a benchmark and to illustrate the effect of the interference limited region associated with turbo codes, coded single user performance is firstly considered on the AWGN channel (Section 8.6.1) and followed by CDMA CDTD/TDTD multiuser performance (Section 8.6.2) on a 2-path Rayleigh channel.

8.6.1 AWGN Performance

Single user performance comparisons between convolutional and turbo coding under identical complexity constraint requirements is carried out, by using (8.7) and (8.7). Figure 8.6 depicts the convolutional and turbo code bounds on the AWGN channel. For the turbo code performance, both the Divsalar (unconditioned) and binomial (conditioned) CPDFs have been used in the calculation of $P_d(\mathbf{c} \to \hat{\mathbf{c}})$ given in (8.7). As expected, a tighter bound is achieved by the binomial CPDF. From Figure 8.6, it is noted that the union bound (using the Divsalar CPDF) for the $R_c = 1/2$ turbo code, diverges at low values of E_b/N_o, for all N_{tc}. Consistent with the results by Divsaler et al. [55], the divergence occurs roughly when the SNR (E_b/N_0) falls below the threshold determined by the computational cutoff rate $R_o{}^2$.

From Figure 8.6 it is clear that the conditioned CPDF results in an improved bound, with a slight divergence around the cutoff rate threshold. As an alternative the actual performance curve can be upper bounded by the uncoded binary PSK performance [205].

The turbo code performance reflects the expected interleaver gain in the waterfall region[3] of the performance curve. This provides an effective way to decrease the BER without invoking any changes in the system configuration.

Figure 8.7 illustrates the performance of low rate convolutional- and turbo codes for a fixed interleaver size, $N_{tc} = 256$. As expected, it is observed that turbo code performance is superior to convolutional code performance for low values of E_b/N_o (< 5 dB). It is noted that for a fixed interleaver size, decreasing the code rate does not significantly affect the waterfall performance region. Decreasing the code rate, however, causes the error floor region to be lowered which may be attributed to the stronger code structure of the low rate codes.

SPACE-TIME CODED TRANSMIT DIVERSITY FOR CDMA 203

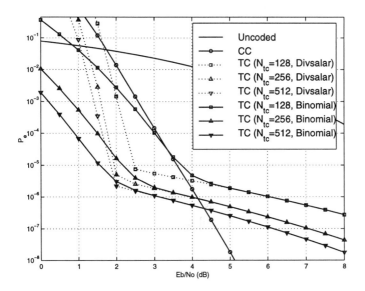

Figure 8.6. CC and TC bounds on AWGN channel with $R_c = 1/2$ and $K = 1$, as a function of turbo interleaver size, N_{tc}.

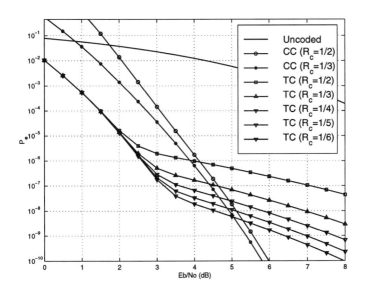

Figure 8.7. CC and TC bounds on AWGN channel with $N_{tc} = 256$ and $K = 1$, as a function of code rate.

Since this region occurs at higher SNR values, the actual weight spectrum becomes more important in influencing the performance.

Since the performance of a single user matched filter CDMA is interference limited, the uncoded BER region of importance for a turbo code is roughly $10^{-4} < P_e < 10^{-2}$. This is the focus BER range and coding should provide acceptable performance in this region.

As another means to investigate multiuser performance, let us define the system load as the quantity $V = K/N_{tot}$, where $N_{tot} = N \cdot R_c$. The system load is therefore the number of active users normalized to the overall spreading factor. Figure 8.8 depicts the system load for low rate convolutional- and turbo coding. It is clear that the system load using turbo codes are substantially higher than for convolutional coding. Another interesting effect is that the coding gain for low rate turbo codes are reduced as the system load increases. This is due to the error floor effect as seen in Figures 8.6 and 8.7. It should be noted that low rate turbo coding will only exhibit this behavior at relatively high E_b/N_0. If the system is operated at low E_b/N_0, low rate turbo coding provides an increase in system capacity.

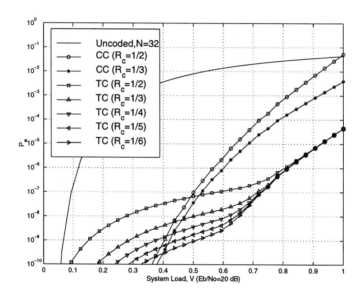

Figure 8.8. Analytical AWGN system load V – low rate convolutional and turbo coding ($N_{tc} = 256$).

8.6.2 Fading Channel Multiuser Performance

Using (8.21), (8.15) and (8.23), Figures 8.9 and 8.10 depict the BER performance of Orthogonal CDTD (O-CDTD) 2-path Rayleigh fast- and slow fading respectively, with rate $R_c = 1/2$ convolutional and turbo ($N_{tc} = 256, 2048$)

SPACE-TIME CODED TRANSMIT DIVERSITY FOR CDMA 205

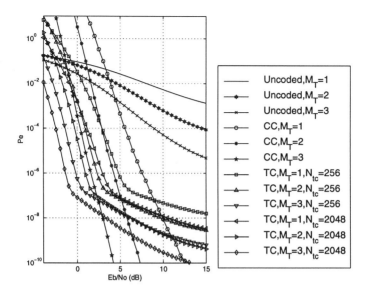

Figure 8.9. Analytical BER performance of coded O-CDTD, with $R_c = 1/2$, $K = 5$, and $M_T = 1, 2, 3$, on a fast fading 2-path channel.

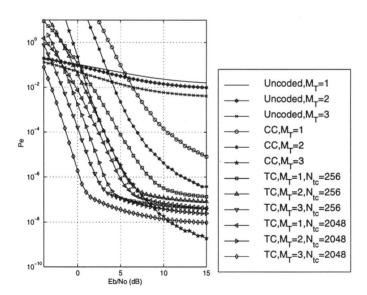

Figure 8.10. Analytical BER performance of coded O-CDTD, with $R_c = 1/2$, $K = 5$, and $M_T = 1, 2, 3$, on a slow fading 2-path channel.

coding. For both graphs, the number of users $K = 5$ with $M_T = 1, 2$ and 3 transmit antenna elements.

By introducing multiple transmit antennas, the diversity order is increased, and thereby the probability of coding gain is increased. This is especially true for turbo coded transmit diversity which is better suited for Gaussian-like MAI. From the graphs it clear that turbo coded transmit diversity increases the performance substantially.

Using the bounds on BER performance derived in Section 8.5, Figures 8.11 to 8.15 depict the performance as a function of system load, V for coded O-CDTD CDMA. The analysis is restricted to a fully interleaved (i.e., fast fading) two path Rayleigh fading channel. The operating point has been taken as $E_b/N_0 = 20$ dB. Figures 8.11 and 8.12 compare coded ($R_c = 1/2$) and uncoded O-CDTD BER performance as a function of interleaver size and transmit diversity order, with $M_T = 1, 2$. Results for single transmit diversity ($M_T = 1$) are included in an attempt to isolate the performance improvement achieved with temporal diversity from the total space-time diversity gain. From the curves it is clear that the introduction of coded transmit diversity has improved the capacity of all the coded systems substantially. For regions of high system load, the best performance is achieved by the turbo coded systems.

Figures 8.13 and 8.14 compare the performance of different low code rates and transmit diversity order, $M_T = 1, 2$. For the turbo coded systems the interleaver size was 256. The low rate codes provide improved performance over the complete range. This illustrates the effectiveness of low rate codes to overcome loss in processing gain under equal bandwidth conditions.

Figure 8.15 compares the performance of coded O-CDTD with $R_c = 1/2$ and transmit diversity order, $M_T = 1, 3$. From the graphs it can be seen that the best performance is always achieved with the highest order diversity.

8.6.3 CDTD and TDTD Comparison

Here we compare the BER performance of orthogonal CDTD (O-CDTD) and optimum antenna-selection TDTD (AS-TDTD). In Figure 8.16, the performance of these transmit diversity schemes with $R_c = 1/2$ and $M_T = 3$ is compared.

It is expected that O-CDTD should outperform AS-TDTD, provided that the orthogonality and conditions of statistical independence are not compromised. From the performance curves it is noted that this argument is correct for system loads less than $V = 0.5$. When the system load is in excess of $V = 0.5$, AS-TDTD signalling outperforms O-CDTD. This can be attributed to the increase in MAI, since for O-CDTD the effective number of simultaneous channels is increased from K to $K \cdot M_T$.

8.6.4 Effects of Fading Correlation

Here we consider the effect of fading correlation on the BER performance of coded O-CDTD. It has been argued that maximum theoretical capacity (or

SPACE-TIME CODED TRANSMIT DIVERSITY FOR CDMA 207

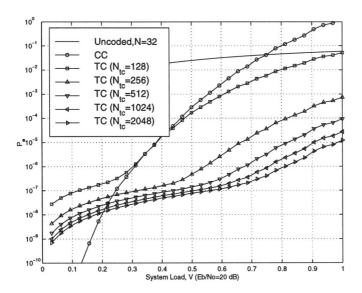

Figure 8.11. Coded O-CDTD BER performance on a 2-path fading channel with $R_c = 1/2$ and $M_T = 1$.

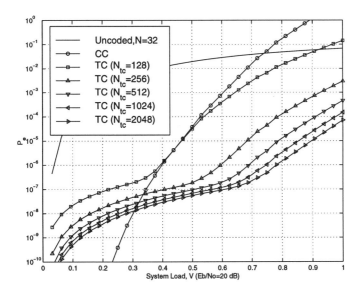

Figure 8.12. Coded O-CDTD BER performance on a 2-path fading channel with $R_c = 1/2$ and $M_T = 2$.

208   SPACE-TIME PROCESSING FOR CDMA

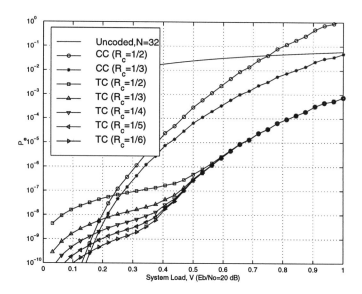

Figure 8.13. Coded O-CDTD BER performance on a 2-path fading channel with $N_{tc} = 256$ and $M_T = 1$.

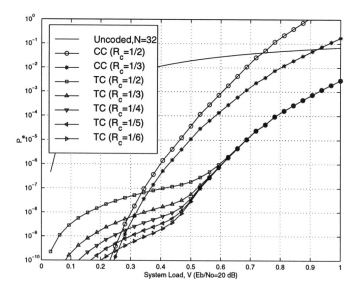

Figure 8.14. Coded O-CDTD BER performance on a 2-path fading channel with $N_{tc} = 256$ and $M_T = 2$.

SPACE-TIME CODED TRANSMIT DIVERSITY FOR CDMA 209

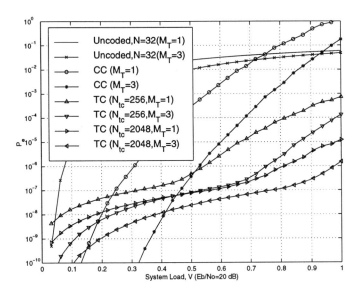

Figure 8.15. Coded O-CDTD performance comparison with $R_c = 1/2$, $N_{tc} = 256, 2048$ and $M_T = 1, 3$.

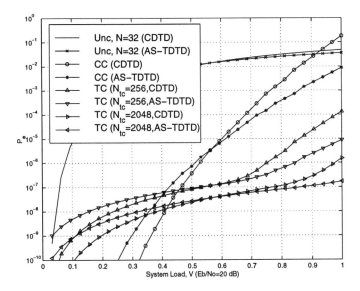

Figure 8.16. Comparison of $R_c = 1/2$ convolutional and turbo coding ($N_{tc} = 256, 2048$) CDTD and AS-TDTD.

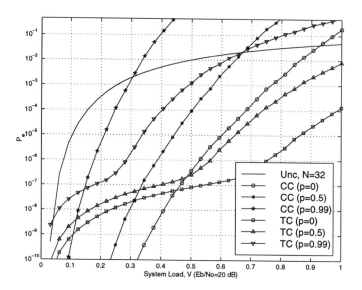

Figure 8.17. Comparison of $R_c = 1/2$ convolutional and turbo coding ($N_{tc} = 256, 2048$), with $M_T = 3$ and $\rho = 0, 0.5, 0.99$.

diversity advantage) can be achieved if uncorrelated signals are transmitted, i.e., the fading experienced by each transmit-receive path is statistically independent.

As explained in Chapter 3, we can have transmit antennas (at the base station) which are correlated. In order to have uncorrelated transmit diversity, we have to separate the antenna elements far apart ($\approx 40\lambda$), which may not be practical.

To investigate the influence of correlation, Figure 8.17 depicts the O-CDTD performance with $R_c = 1/2$, $M_T = 3$, and fading correlation coefficients, $\rho = 0.0, 0.5$ and 0.99. The performance degradation due to correlation is not that significant if ρ is restricted to 0.5. However, the performance is severely degraded when $\rho \geq 0.5$. The latter can be attributed to the fact that the correlated multi-path channel has memory which reduces the effectiveness of the combined coding/diversity scheme. It should be noted that, since ideal channel estimation is assumed, the results do not reflect the additional performance degradation associated with imperfect channel estimation. Under conditions of correlated fading, strong emphasis should be placed on the channel estimator in an attempt not to degrade system performance even further.

Figure 8.18 shows the O-CDTD performance in a NLOS environment. Correlation values are for the Gaussian scattering environment described in Section 4.2.2. At low system loads the higher order O-CDTD has superior performance. At higher loads, however, the MAI increases and no gain for a higher order O-CDTD scheme is possible. Correlation has a more pronounced degrading effect for higher transmit diversity order. These conclusions are consistent for both convolutional- and turbo codes, with turbo codes having consistently better performance than convolutional codes.

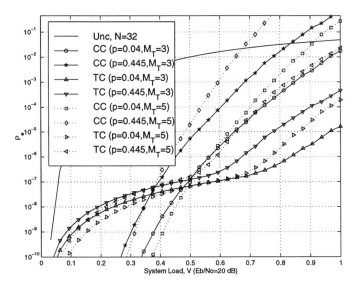

Figure 8.18. O-CDTD NLOS performance with $M_T = 3, 5$.

8.7 SIMULATION PERFORMANCE

Using the MATLAB based UMTS-like WCDMA Simulation Environment described in Appendices H and G, and system parameters as outlined in Table 8.1, this section discusses WCDMA related coded space-time performance results. As with the preceding analytical results, space-time convolutional and turbo coded CDTD and TDTD are considered. In the simulations the vehicular channel is considered at a vehicle speed of 120 km/h. RAKE receivers with $L_R = 3$ fingers are used. It is assumed that these RAKE receivers have perfect knowledge of the excess delays on each multi-path component.

Figure 8.19 illustrates the performance of O-CDTD with convolutional and turbo coding in a multiuser WCDMA system. For a fair comparison between no transmit diversity and $M_T = 2$ transmit diversity, the signal power transmitted per antenna has been reduced (halved) in order to ensure the same radiated power as for the single transmit antenna system. For the same order of diversity, turbo coding out performs convolutional coding and reduces the error performance to an acceptable level for speech.

It is also important to note that even when employing orthogonal spreading sequences (as in O-CDTD), the downlink will not be perfectly orthogonal due to multi-path propagation. In [281, 279], the downlink orthogonality factor has been calculated for different environments. This factor, expressed as a per-

212 SPACE-TIME PROCESSING FOR CDMA

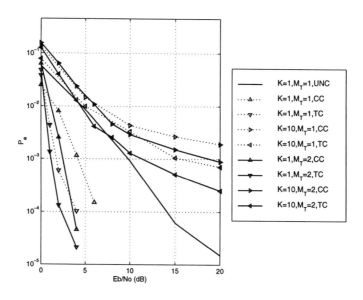

Figure 8.19. Average error rate for O-CDTD in the WCDMA vehicular environment (120 km/h).

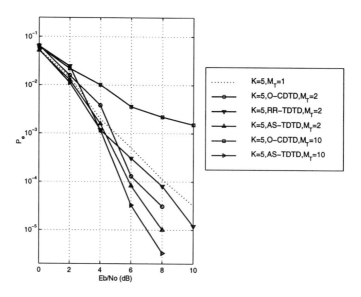

Figure 8.20. Average error rate for different transmit diversity schemes in the WCDMA vehicular environment (120 km/h).

Parameter		Simulation value
Spreading sequence length		$N = 32$
Operating environment		Vehicular channel (120 km/h)
Temporal fading		$m = 1$ (NLOS micro-cell)
Multi-path signals		$L = 3$
Users		$K = 1, 5, 10$
RAKE fingers		$L_R = 3$
RAKE receivers	O-CDTD	M_T
	AS-TDTD	1
	RR-TDTD	1
Antennas	Transmitting	$M_T = 1, 2, 10$
	Receiving	$M_D = 1$
FEC code rate		$R_c = 1/3$

Table 8.1. System parameters for simulation performance.

centage, is the fraction of the total output power that will be experienced as intra-cell interference. An orthogonality factor of zero corresponds to a perfectly orthogonal downlink, while a factor of one is a completely non-orthogonal downlink. As shown in Table 8.2, 40% of the power transmitted from the reference cell will act as intra-cell interference in a vehicular environment.

Propagation model	Orthogonality factor (%)
Indoor office	10 %
Outdoor to indoor and pedestrian	6 %
Vehicular	40 %

Table 8.2. Orthogonality factor for the different WCDMA channel environments.

Orthogonality, therefore, removes 60% of the interference or, stated differently, an orthogonality factor of 0.4 is obtained (40% of the interference remains). This observation is important since it provides an explanation for the sub-optimum performance obtained with O-CDTD. Under these conditions, NO-CDTD should provide improved performance.

In Figure 8.20, the performance of convolutional coded O-CDTD, RR-TDTD and AS-TDTD CDMA, with $M_T = 2, 10$, are considered on the vehicular channel. For both the RR-TDTD and AS-TDTD schemes, the transmitting antenna is switched in a deterministic manner after every slot. For the RR-TDTD scenario the selection is based on a pseudo-random (round-robin) antenna hopping sequence. In the AS-TDTD implementation, the transmit antenna selection is performed more optimally. Here, closed-loop antenna selection (CL-AS) from every mobile receiver to base station transmitter is used.

As the number of antennas is increased to 10, the performance of O-CDTD is severely degraded due to the increased level of MAI. With this high transmit diversity order the best performance is achieved with AS-TDTD. However, the performance improvement is not substantial. This can be attributed to the fact that since antenna selection is performed, the selected transmission path remains in a fading state. For this reason only selection diversity performance gains can be expected.

8.8 SPACE-TIME CODE EXTENSIONS

Two extensions to layered space-time coding is presented in this section. These are turbo transmit diversity (TTD) [288, 291, 286] and super-orthogonal turbo transmit diversity (SOTTD) [289, 287].

TTD differs from other conventional diversity combining techniques in that the original rate $R_c = 1/2$ code is transformed into a more powerful rate $R_c = 1/(Z+1)$ code. The way the turbo coder is configured fits naturally into the transmit diversity schemes described above.

8.8.1 Turbo Transmit Diversity (TTD)

In TTD the basic principles of CDTD is extended to include low rate turbo coding for orthogonal or non-orthogonal CDTD, and is an extension of the work by Barbalescu [18]. The principle of operation is to transmit the coded bits, stemming from the constituent RSC encoders, via the spatial domain rather than via the time, code or frequency domain. The received data stream is then iteratively decoded using turbo decoding principles. The power of TTD lies in the principle that a single-antenna rate $R_c = 1/2$ coder is transformed into a more powerful turbo code with rate $R_c = 1/(Z+1)$, where Z is the number of constituent encoders as indicated in Figure 8.21.

The TTD scheme is based on three underlying principles. These are the

- rate $R_c = 1/(Z+1)$ turbo encoding,
- puncturing and multiplexing, and
- the iterative decoding.

These principles are described in detail below.

8.8.1.1 Rate-$1/(Z+1)$ Turbo Encoding. The constituent turbo encoder of Figure 8.21 produces one uncoded output systematic stream $\mathbf{x}_0^{(j)}$ and Z encoded output parity streams, denoted by $\mathbf{x}_1^{(j)}, \ldots, \mathbf{x}_Z^{(j)}$. Here

$$\mathbf{x}_Z^{(j)} = \{x_{0,z}^{(j)}, x_{1,z}^{(j)}, x_{2,z}^{(j)}, \ldots, x_{i,z}^{(j)}, \ldots, x_{(N_{tc}-1),z}^{(j)}\}, \; (z = 0, 1, \ldots, Z), \qquad (8.24)$$

SPACE-TIME CODED TRANSMIT DIVERSITY FOR CDMA 215

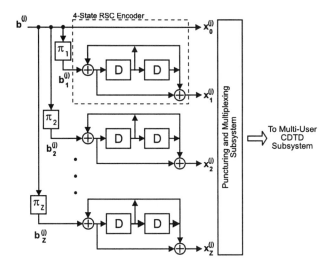

Figure 8.21. Generalized TTD block diagram.

where i is the discretized time index, and j denotes the reference user. The parity streams are produced by Z identical RSC encoders with constraint length K_{tc}. In the discussion that follows the constrained length $K_{tc} = 3$.

The first component encoder operates directly (or through interleaver π_1) on the information sequence, $\mathbf{b}_1^{(j)}$ of length N_{tc}, producing two output sequences $\mathbf{x}_0^{(j)}$ and $\mathbf{x}_1^{(j)}$. The second component encoder operates on a re-ordered sequence of information bits, $\mathbf{b}_2^{(j)}$, produced by interleaver π_2, also of length N_{tc}, and outputs the sequence $\mathbf{x}_2^{(j)}$. The systematic information bit stream of this RSC encoder is discarded. Subsequent component encoders operate on a re-ordered sequence of information bits, $\mathbf{b}_Z^{(j)}$, produced by interleaver π_Z, and output the sequence $\mathbf{x}_Z^{(j)}$.

8.8.1.2 Puncturing and Multiplexing. The puncturing and multiplexing procedures form the heart of TTD. As an example, to show how a rate $R_c = 1/2$ code is being transformed into a rate $R_c = 1/(Z+1)$ code by appropriate puncturing and multiplexing, consider a turbo encoder with Z constituent RSC encoders and a single transmit antenna $M_T = 1$. Assuming further for our example QPSK modulation, the information and coded sequences of user j can be arranged in terms of the in-phase and quadrature phase components. The in-phase component of the QPSK modulator (I branch) transmits the systematic bits, while the quadrature component (Q branch) transmits the parity bits formed by the constituent encoders. Beginning at discrete time $i = 0$, the in-phase component is modulated by

$$I: \left\{ x_{0,0}^{(j)}, x_{1,0}^{(j)}, \ldots, x_{(Z-1),0}^{(j)}, x_{Z,0}^{(j)}, x_{(Z+1),0}^{(j)}, \ldots, x_{(N_{tc}-2),0}^{(j)}, x_{(N_{tc}-1),0}^{(j)} \right\}, \quad (8.25)$$

while the quadrature component is modulated by

$$Q: \left\{ x_{0,1}^{(j)}, x_{1,2}^{(j)}, \ldots, x_{(Z-1),Z}^{(j)}, x_{Z,(Z+1)}^{(j)}, x_{(Z+1),(Z+2)}^{(j)}, \ldots, \right.$$
$$\left. x_{(N_{tc}-2),(Z-1)}^{(j)}, x_{(N_{tc}-1),Z}^{(j)} \right\}. \quad (8.26)$$

This puncturing and multiplexing procedure is illustrated in Figure 8.22(a), using the notation $x_{i,Z}^{(j)}$. It should be pointed out that, due to the puncturing procedure, some of the coded sequences are not transmitted to maintain the $R_c = 1/2$ coding rate.

The single transmit antenna $M_T = 1$ example can easily be extended to $M_T = Z$ transmit antennas. In TTD with $(M_T > 1)$, the systematic information sequences are repeatedly transmitted on the I branches of all the available transmit antennas. This is done to guarantee soft failure in order to achieve maximum space-time diversity gain. In addition, smart puncturing and multiplexing is employed to assign the parity information sequences to the different Q branches available for transmission. This puncturing and multiplexing procedure is shown in Figure 8.22(b) for $M_T = Z = 3$. Note that the information and coded sequences transmitted by the I and Q branches of the first antenna element for $M_T > 1$, agrees with the single antenna transmission of Figure 8.22(a).

The I and Q sequences transmitted by the second antenna can be written as

$$I: \left\{ x_{0,0}^{(j)}, x_{1,0}^{(j)}, \ldots, x_{Z,0}^{(j)}, x_{(Z+1),0}^{(j)}, \ldots \right\}$$
$$Q: \left\{ x_{0,2}^{(j)}, x_{1,3}^{(j)}, \ldots, x_{(Z-2),Z}^{(j)}, x_{(Z-1),1}^{(j)}, x_{Z,2}^{(j)}, \ldots \right\}. \quad (8.27)$$

In general, the sequences transmitted by the zth antenna element can be written as

$$I: \left\{ x_{0,0}^{(j)}, x_{1,0}^{(j)}, x_{2,0}^{(j)}, x_{3,0}^{(j)}, x_{4,0}^{(j)}, x_{5,0}^{(j)}, \ldots \right\}$$
$$Q: \left\{ x_{0,z}^{(j)}, x_{1,(z+1)}^{(j)}, \ldots, x_{(Z-z-1),Z}^{(j)}, x_{(Z-z),1}^{(j)}, \ldots \right\}. \quad (8.28)$$

It should be noted that none of the encoded information bits are lost by the puncturing and multiplexing operations, while the effective transmission per constituent QPSK transmission rate remains one half.

8.8.1.3 Iterative Decoding.
Figure 8.23 depicts the iterative decoding configuration. Before the decoding is performed, the demodulated signal streams

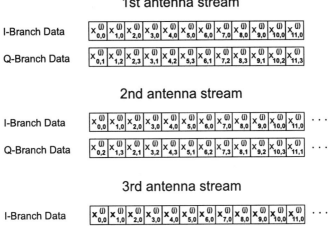

Figure 8.22. Puncturing and multiplexing procedure for a rate $R_c = 1/2$ turbo encoder with Z constituent RSC encoders. (a) Single transmit antenna, $M_T = 1$ (b) $M_T = 3$ transmit antennas.

are de-multiplexed. For the punctured symbols values are obtained from the $Z - 1$ received antenna streams. For single antenna TTD, zero values are inserted in the punctured bit positions. The decoder therefore regards the punctured bits as erasures.

From our previous single antenna TTD example, the following sequences are inputs to the zth decoder

$$I: \quad \left\{ y_{0,0}^{(j)}, y_{1,0}^{(j)}, \ldots, y_{Z,0}^{(j)}, y_{(Z+1),0}^{(j)}, \cdots \right\}$$
$$Q: \quad \left\{ y_{0,z}^{(j)}, 0, 0, \ldots, 0, y_{Z,z}^{(j)}, \cdots \right\}. \tag{8.29}$$

The iterative decoding procedure requires Z component decoders using soft inputs and providing soft outputs, based on the MAP algorithm. The decoding configuration operates in *serial mode*, i.e. decoder 1 processes data before decoder 2 starts its operation, and so on [52]. Many different configurations exist, especially for the case where $Z \geq 3$.

With reference to Figure 8.23, extrinsic information (related to a data symbol) is obtained from surrounding symbols in the codeword sequence imposed by the code constraints only. The extrinsic information is obtained without any information concerning the symbol itself, and is provided as soft outputs by the component decoders. The *soft outputs*, obtained from the MAP, are internal variables of the decoder, and is a measure of the reliability of the decoding of single bits and do not provide hard bit decisions.

In addition, intrinsic information related to a data symbol is *a priori* information attached to the symbol without using any code constraints. This information is used by the component decoders as additional information related to each code symbol. In iterative decoding, the extrinsic information provided by the previous decoding step becomes the *a priori* information of the current decoding process.

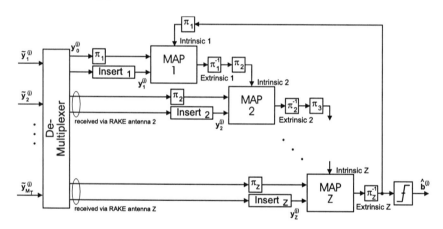

Figure 8.23. Generalized block diagram of a serial iterative turbo decoder.

8.8.2 Super-orthogonal Turbo Transmit Diversity (SOTTD)

To provide a more flexible architecture for the generation of variable low rate coding, SOTTD extends TTD to include Z component decoders, and is roughly based in the work by Pehkonen et al. [196, 197].

The general structure of the SOTDD encoding scheme is illustrated in Figure 8.24. The SOTDD encoding scheme is formed by two rate $R_c = 2/32$ encoders, consisting of a rate $R_c = 2/5$ RSC encoder and a rate $R_c = 5/32$ Walsh Hadamard (WH) orthogonal modulator, denoted by (RSC&WH). These encoders are parallel concatenated through an interleaver. The first encoder processes the original data sequence, whereas before passing through the second

SPACE-TIME CODED TRANSMIT DIVERSITY FOR CDMA 219

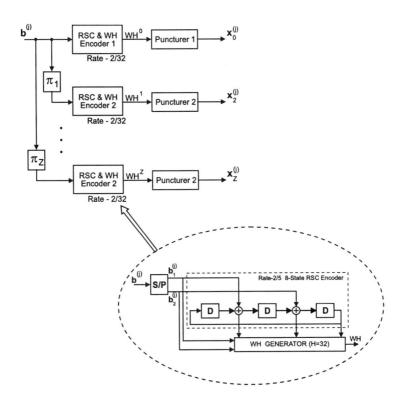

Figure 8.24. SOTTD block diagram.

encoder, the data sequence is permuted by a *pairwise* pseudo random interleaver of length N_{tc}. The outputs of the two constituent encoders are punctured in order to provide a wide range of code rates. These outputs are then punctured and multiplied as described in Section 8.8.1.2.

The detailed structure of the combined RSC turbo and WH encoder is also shown in Figure 8.24. The state outputs of the rate $R_c = 2/5$ RSC encoder is fed to the rate $R_c = 5/32$ WH, producing one sequence of length ($L_{WH} = 32$) from a set of $H = 5^2 = 32$ sequences.

It is well known that the most important characteristic of a codeword is its minimum free distance [223]. Owing to the orthogonality characteristics of the WH codewords, the minimum distance of the encoded sequences for both constituent encoders is equal to $L_{WH}/2 \cdot N_{tc} = 16\, N_{tc}$.

After encoding, the output sequences are obtained by appropriate puncturing according to puncturing patterns $P^{(i)} = \{p_1^i, p_2^i, \ldots, p_{N_{tc}}^i\}$, where $i = 1$ and

$i = 2$ for the first and second puncturer, respectively. With $W_{P(1)}$ and $W_{P(2)}$, respectively the weights of the first and second puncturers, the resulting overall encoder rate is given by

$$R_c = \frac{2}{W_{P(1)} + W_{P(2)}}. \tag{8.30}$$

Therefore, when none of the output sequence bits are punctured, the overall code rate of the combined turbo and WH encoding is $R_c = 2/(32+32) = 2/64$.

The parallel concatenated structure of the turbo code permits it to be decoded in parts by two or more decoders. For this purpose the different SISO algorithms available for turbo decoding may be considered. A detailed description of the decoder configuration is not presented here, but the interested reader is referred to [196, 197, 289, 287].

8.9 SUMMARY

This chapter has considered space-time coded transmit diversity techniques as a means to improve cellular CDMA performance. The suitability of convolutional- and turbo coding, when applied to layered space-time transmit diversity, was discussed and analytical results presented for CDTD and TDTD in AWGN and fading channels. Analytical results have shown that combining spatial and temporal processing at the transmitter (the downlink) is an effective way to increase CDMA system capacity. Link-level simulations based on the UMTS-like WCDMA simulation environment have also shown the power of space-time transmit diversity in a more realistic environment.

From the analytical and simulation results it was shown that turbo coding, when applied to space-time transmit diversity, outperforms convolutional codes with the same code rate and comparable complexity. In general, the average performance of the CDTD and TDTD is comparable.

Extensions of CDTD have been presented in the form of TTD and SOTTD. These schemes have the ability to improve the cellular capacity even further.

Notes

1. In an fully interleaved scheme, fast fading is created as the de-interleaving mechanism creates a virtually memoryless channel [37].

2. The cutoff rate is defined as $E_s/N_o = R_c E_b/N_o < -\ln(2^{1-R_c} - 1)$ for a code with rate R_c [205].

3. The "waterfall region" is defined, as the part of the performance curve where the BER decreases rapidly with increased SNR. The region where the BER performance changes very slowly, with increased SNR, is defined as the "error floor region".

Appendix A
List of Abbreviations

2G	second generation
3G	third generation
ACTS	advanced communication technology system
ADSL	asymmetric digital subscriber loops
AGC	automatic gain control
AMPS	advanced mobile phone system
APON	ATM passive optical network
AS-TDTD	antenna-selection time-division transmit diversity
ATM	asynchronous transfer mode
AWGN	additive white Gaussian noise
BEP	analytical bit error probability
BER	bit error rate
B-ISDN	broadband ISDN
b-JOR	block Jacobi over relaxation
BPSK	binary phase-shift keying
BS-CMA	beam space CMA
BSS	base station subsystem
BU	bad urban
CCPCH	common control physical channels
CDS	circular disk of scatterers
CDMA	code-division multiple-access
CDTD	code-division transmit diversity
CL-AS	closed-loop antenna-selection
CLPC	closed-loop power control
CMA	constant modulus algorithm

CPDF	conditional probability distribution function
CSD	clipped soft decision
CSI	channel state (or side) information
CT-2	cordless telephone -2
CTM	cordless telephone mobility
dB	decibel
DCS 1800	digital communication system - 1800
DD	decision-directed
DDFD	decorrelating decision feedback detector
DECT	digital enhanced cordless telephone
DOA	direction-of-arrival
DPCCH	dedicated physical control channel
DPCH	downlink physical channel
DPDCH	dedicated physical data channel
DQPSK	differential quadrature phase shift keying
DS	direct-sequence
EC	European commission
EFD	effective fading distribution
EGC	equal gain combining
EL-CMA	element space CMA
EMF	estimated matched filter
ERC	European radio committee
ESM	effective scatterer model
ESPRIT	estimation of signal parameters via rotational invariance technique
ETSI	European telecommunications standards institute
EU	European union
FDD	frequency-division duplex
FDMA	frequency-division multiple-access
FEC	forward error correction
FFT	fast Fourier transform
FIR	finite impulse response
FM	frequency modulation
FMA2	FRAMES multiple access mode 2
FPLMTS	future personal land mobile telephone system
FRAMES	future radio wideband multiple access systems
FSM	finite state machine
FTF	fast transversal filter
FTTB	fibre to the building
FTTC	fibre to the curb
FTTH	fibre to the home

GAA	Gaussian angle of arrival (model)
GBSBEM	geometrically based single bounce elliptical model
GFD	Gaussian fading distribution
GS	Gauss Seidel
GSM	global system for mobile communications
HD	hard decision
HDSL	home digital subscriber line
HFC	hybrid fibre copper
HFR	hybrid fibre radio
HSR	high sensitivity reception
IC	interference cancellation
IIR	infinite impulse response
ILSE	iterative least squares enumeration
ILSP	iterative least squares projection
IMT-2000	international mobile telecommunications for the 21st century
IN	intelligent network
IP	internet protocol
IPI	inter-path-interference
ISDN	integrated services digital network
ISI	inter-symbol interference
IS-54	interim standard 54
IS-95	interim standard 95
ITU	international telecommunications union
JOR	Jacobi over relaxation
LAN	local area network
LEO	low earth orbit (satellite)
LMDS	local multi-point distributed system
LMS	least mean squares
LMMSE	least mean minimum square error
LOS	line of sight
LS	least squares
LS-DD	least squares decision directed
LS-DRMTA	least squares despread respread multi target array
LS-DRMTCMA	least squares despread respread multi target CMA

LSED	layered squared Euclidean distance
LSEDP	layered squared Euclidean distance product
LSL	least square lattice
MAC	multiple-access channel
MAI	multiple-access interference
MAP	maximum a posteriori
MBS	mobile broadband systems
MF	matched filter
MIMO	multiple input/multiple output
MIP	multi-path intensity profile
ML	maximum likelihood
MLSD	maximum likelihood sequence detection
MM	multi media
MMDS	microwave multi-point distributed radio
MM-LEO	multi media LEO
MMSE	minimum mean square error
MOU	memorandum of understanding
MRC	maximal ratio combining
MSE	mean square error
MTCM	multiple trellis-coded modulation
MT-DD	multi target DD
MT-LSCMA	multi target least squares CMA
MU	multiuser
MUD	multiuser detection (or demodulation)
MUSIC	multiple signal classification
NLMS	normalized LMS
NLOS	non-line of sight
NSS	network switching subsystem
O-CDTD	orthogonal code-division transmit diversity
O-CMA	orthogonalized CMA
OVSF	orthogonal variable spreading factor
PABX	private automatic branch exchange
PC	personal computer
PCS 1900	personal communication system 1900
PIC	parallel interference cancellation
PDF	probability distribution function
PDNW	partially decorrelating noise whitening

PHS	personal handi-phone system
PN	pseudo-noise
PRACH	physical radio access channel
PSK	phase-shift keying
QoS	quality-of-service
QOSI	quality of service inhomogeniety
QPSK	quadrature phase-shift keying
RACE	research on advanced communications in Europe
R-CMA	recursive CMA
RF	radio frequency
RMS	root mean square
RLS	recursive least squares
RLSE	recursive least squares enumeration
RSC	recursive systematic convolutional
RR-TDTD	round-robin time-division transmit diversity
SCH	synchronization channel
SCORE	self-coherence restoral (algorithm)
SD	soft-decision
SD-DD	steepest decend DD
SDH	subscriber digital hierargy
SDMA	space division multiple access
SFIR	spatial filtering for interference rejection
SIC	successive (or serial) interference cancellation
SIM	subscriber identification module
SINR	signal-to-interference-noise ratio
SISO	single (soft) input/single (soft) output
S-LEO	satellite LEO
SMG	special mobile group
SNR	signal-to-noise ratio
SOTTD	super-orthogonal turbo transmit diversity
SOVA	soft-output Viterbi algorithm
SPD	semi positive definite
SS	spread-spectrum
STCM	space-time coded modulation
SU	single user
TD-CDMA	time division CDMA
TDD	time-division duplex
TDMA	time-division multiple-access

TDTD	time-division transmit diversity
T/F/C	time, frequency and code
TFI	transport-format indicator
TOA	time of arrival
TPC	transmit power control
TSUNAMI	technologies in smart antennas for universal advanced mobile infrastructure
TTD	transmit turbo-diversity
TU	typical urban
UHF	ultra high frequency
ULA	uniform linear array
UMTS	universal mobile telephone system
UTRA	UMTS terrestrial radio access
VHE	virtual home environment
VPL	vertical plane launch
V-SAT	very small apperture antenna terminal
VSF	variable spreading factor
WATM	wireless ATM
WLAN	wireless local area network
WLL	wireless local loop
WSS	wide-sense stationary
WCDMA	wideband code-division multiple-access
xDSL	high speed digital subscriber line

Appendix B
List of Symbols

arg	argument
max	maximum
var $\{\mathbf{y}\}$	variance of $\mathbf{y}$
$\{-1, 1\}$	binary set
$\|\cdot\|$	Euclidean distance
$(\cdot)^{\mathrm{H}}$	hermitian
α_{cma}	amplitude of array output
$\mathbf{A}$	matrix for design of linear cancellation
$A_{cluster}$	the physical area occupied by the re-use cluster
A_d	number of codewords of weight d
$A_{gff/gfb}$	feedforward and feedback generator polynomial
A_k	received amplitude
$a_k(t)$	binary spreading waveform
A_{norm}	normalizing factor
$AF(\phi, \theta)$	array factor
β_k	received signal strength in volts
$\beta_l^{(k)}(i)$	strength of specific multi-path
B_e	blocking rate
$\mathbf{b}$	vector of data bits
$\hat{\mathbf{b}}$	specific version of $\mathbf{b}$
$\mathbf{b}_i$	vector of bits for interval i
$b_k(t)$	binary data sequence
$b_i^{(k)}$	bit for user k, symbol interval i
$\tilde{\mathbf{b}}_i^{(k)}$	vector of bits up until user k symbol interval i

$\hat{\mathbf{b}}(s)$	bit estimate vector for stage s
$\hat{\mathbf{b}}_{Dec}$	estimate from the decorrelator
$\hat{\mathbf{b}}_{ML}$	maximum likelihood estimate of $\mathbf{b}$
$\hat{\mathbf{b}}_{MMSE}$	estimate from the MMSE
$\mathcal{BW}$	total signal bandwidth
$\mathcal{BW}_{ch}$	coherence bandwidth
$\mathbf{C}$	complex fading matrix
$\mathbf{C}_i$	complex fading matrix for symbol interval i
C_{tot}	total capacity
c	speed of light
$c_l^{(k)}(i)$	complex fading coefficient for specific multi-path of symbol i
$\hat{c}_l^{(k)}(t)$	complex fading process for specific multi-path
$\mathbf{c}_n$	codeword
$\mathbb{C}^{(X)}$	X dimensional complex space
$\mathbb{C}^{X \times Y}$	$X \times Y$ dimensional complex space
$\mathbb{C}$	set of complex numbers
$\mathbf{D}$	diagonal matrix of $\mathbf{R}$
$\mathbf{D}_n$	diagonal of $\mathbf{D}_{n,n}$
$\mathbf{D}_{n,n}$	diagonal matrix for interval n
d	Hamming weight
d_{clus}	number of scattering clusters
d_{free}	free distance
$d_i^{(k)}$	element for user k symbol interval i of $\mathbf{D}$
$d^{(j)}(t)$	reference signal for user j
d_x	distance between antenna elements
Δ	maximum angular spread of signals arriving at base station
Δf	frequency separation
$\delta[\cdot]$	delta function
δ_d	rate of average power decay
δ_m	exponential fading distribution parameter
$\text{erfc}(x)$	complimentary error function of x
$\mathrm{E}\{\mathbf{y}\}$	expected value of $\mathbf{y}$
E_b	energy per bit
$E_{d|i}$	expectation w.r.t. distribution $p(d \mid i)$
E_s	energy per symbol
$\mathbf{E}_{\mathbf{total}}(\phi, \theta)$	matrix describing the electrical field of the array
$e_i^{(k)}$	estimation error
η	noise vector for Rake combined case
$\hat{\eta}$	noise vector from decorrelator

APPENDIX B: LIST OF SYMBOLS   229

η_c	cellular spectral efficiency
$\eta_n^{(j)}(i)$	noise sample
η_m	bandwidth efficiency
$\boldsymbol{\eta}_\zeta$	vector of noise samples
$\mathbf{F}$	Cholesky factorisation of $\mathbf{R}$
$\mathbf{f}^{(n)}$	block output vector
f_d	maximum doppler shift
$\mathbf{f}_x(\cdot)$	tentative decision function
$\Phi_S(t)$	characteristic function of S
ϕ_{3dB}	3 dB beam-width of antenna pattern
ϕ_{BW}	maximum beam-width of antenna pattern
ϕ_0	LOS angle
ϕ_j	angle of multi-path component arriving at the mobile with respect to the LOS component
$\phi_l^{(k)}$	broadside angle of arrival for user k
ϕ_ν	angle of mobile motion relative to the LOS component
ϕ_r	angle relative to LOS
ϕ_s	scan angle
$\mathbf{G}_c$	arbitrary combining matrix
$\mathbf{G}_e$	equal gain combining matrix
$\mathbf{G}_R$	Rake combining matrix
$\mathbf{G}_s$	selection diversity combining matrix
$\mathbf{G}_i^{(s)}$	selection diversity combining matrix for symbol interval i
$g(t)$	chip waveform
$\mathbf{g}_e^{(k)}(i)$	equal gain combining vector
g_{ff}	feedforward
g_{fb}	feedback
$\mathbf{g}_R^{(k)}(i)$	Rake combining vector
$\mathbf{g}_s^{(k)}(i)$	selection diversity combining vector
$\Gamma(\cdot)$	Gamma function
Γ_c	signal to noise ratio
γ_b	average SNR per bit
γ_k	average SNR per diversity branch
H	number of codewords
$\mathbf{H}$	MMSE filter
$\hat{\mathbf{h}}_i^{(k)}$	adaptive filter vector for user k symbol interval i
$\mathbf{h}_i^{(k)}$	filter vector for user k symbol interval i
$\mathbf{h}^{(k)}(t,\tau)$	channel impulse response
$h_i^{(k)}(\tau)$	multipath channel response

$\mathbf{I}$	identity matrix
I_{mai}^n	multiple access interference
I_{ni}^n	AWGN interference
I_{si}^n	self interference
i	symbol interval index
ι_x	progressive phase shifts between excitation currents to eleme along the x axis
$J_0(\cdot)$	Bessel function
$J(\hat{\mathbf{h}}_i^{(k)})$	mean squared error for user k symbol interval i
$J_{\text{tot}}(\mathbf{H}^{\text{H}})$	mean squared error
$J(\mathbf{w})$	beamforming cost function
$\nabla J(\mathbf{w})$	gradient of beamforming cost function
j	user index
k	user index
$\mathcal{K}_k$	Rice factor
K	number of active users
K_{subs}	number of subscriber per cell
K_a	length of observation interval
K_{cc}	convolutional code constraint length
K_{tc}	turbo code constraint length
K_{oc}	orthogonal code constraint length
K_0	constant from normal distribution
κ	multiplicity factor (product between number of active users a transmit/receive antennas)
l	multipath index
λ	wave length
$\lambda(\hat{\mathbf{b}})$	log-likelihood metric
$\overline{L}_s(r)$	mean path loss
L_T	path loss threshold in dB
$L_s(r)$	path or free space loss
$\mathbf{L}_n$	lower left of $\mathbf{D}_{n,n}$
$\mathbf{L}_{n,n-1}$	lower left matrix for interval n
$\mathbf{L}$	lower left of $\mathbf{R}$
$L_x(r)$	total path loss
$\overline{L}_x(r)$	mean path loss
L	number of multipath
L_r	number of RAKE taps
L_{min}	length of shortest error event path
L_{WH}	length of Hadamard codeword
μ	step size in the steepest decent algorithm
m_{eff}	effective m after optimum combining

m	Nakagami-m fading parameter
m_0	Nakagami parameter of main received path
M_D	number of diversity antennas
M_B	number of beamforming antennas
M_T	number of transmit antennas
M_C	coder alphabet size
M_{SB}	number of switched-beam antennas
M_{ch}	number of channels per cell
$\mathbf{M}$	off-diagonal part of $\mathbf{R}$
$m_{i,l}^{(kj)}$	element pertaining to user k and j, symbol interval l and i
$\mathbf{m}_i^{(k)}$	element for user k symbol interval i of $\mathbf{M}$
μ_n	step size for block approach
$\mu_i^{(k)}(\tilde{\mathbf{b}}_i^{(k)})$	recursive metric increment
n_l	path loss exponent
n	symbol index
$\mathbf{n}(t)$	Gaussian noise in received signal
$n(t)$	complex envelope of the noise process
$n_m(t)$	Gaussian noise process for antenna element m
N	processing gain
N_B	number of columns in planar array
N_s	number of scatterers
N_{peak}	number of peaks in user distribution
N_{cc}	convolutional code interleaver size (trellis decoder decoding depth)
N_{tc}	turbo code interleaver size (trellis decoder decoding depth)
N_0	two-sided noise spectral density
$N_{S/CH}$	number of subscriber per channel
N_{ch}	number of channels per trunk
N_T	number of channels per re-use cluster
n_{ch}	number of trunks per cell
Ω_k	average power of received path
ω_c	carrier frequency
$p_{R_s,\phi}(R_s,\phi)$	pdf of scatterers
$p_{\Phi_0}(\phi_0)$	pdf of the angular distribution of users
$P(\mathbf{b}|\mathbf{y})$	probability of $\mathbf{b}$ given $\mathbf{y}$
$p(\tau,\phi_r)$	DOA pdf at the base-station at time τ and angle ϕ_r
$p(\mathbf{y}|\mathbf{b})$	pdf of $\mathbf{y}$ given $\mathbf{b}$
$P(\mathbf{b})$	probability of $\mathbf{b}$
$p_S(s)$	pdf of S
P_k	received power
P	number of transmitted symbols

$\mathbf{P}$	cross-correlation matrix
P_e	bit error probability
$P_d(c - \hat{c})$	probability of codeword error
P_w	word error probability
π	interleaver
π^{-1}	de-interleaver
q	multipath index
$Q(\cdot)$	Q-function
r_0	reference distance
r_c	number of cells in a re-use cluster
r_s	distance from base station to arbitrary scattering point
$r_m(t)$	received signal for antenna m
$r_o(t)$	output of antenna array
$\mathbf{r}(t)$	received signal
$\mathcal{R}^{(kj)}$	spatial correlation between user k and reference user
R_{k1}	periodic correlation between user k and reference user
$\hat{R}_{k1}$	aperiodic correlation between user k and reference user
$\mathcal{R}$	data rate
$\mathcal{R}_{ij}$	data rate at some predefined BER available to subscriber i in j of the re-use cluster
$R_D(\Delta t, \phi_r)$	spaced-time correlation function
$\|R_M(\Delta f, \phi_r)\|$	spaced-frequency correlation function
R	mobile to base station distance
R_D	scatterer radius
R_0	cut-off rate of channel
$\mathbf{R}^{-1}$	inverse of $\mathbf{R}$
$R_{jn,jl}^{(x)}(i)$	correlation coefficient for symbol interval i and $i + x$
Re	real part
R_r	radius of cells
R_{xx}	correlation of real components
R_{xy}	correlation between real and imaginary components
$\mathbf{R}$	normalized correlation matrix
$\mathbf{R}_i^{(x)}$	correlation matrix for symbol offset x
$\mathbf{R}_\zeta$	correlation matrix for matched filtered statistic
R_{ij}	correlation between signature waveforms of transmissions i an
$\rho_{jq,kl}^{(x)}(i)$	correlation coefficient for symbol offset x
ρ_{ij}	fading correlation constant between transmissions i and j
ρ_s	average spatial correlation
$\rho(x)$	envelope correlation of signals received x meter apart

APPENDIX B: LIST OF SYMBOLS 233

s	stage index
$s_a(\phi, \theta)$	steering vector
$S_n^{(j)}$	desired received signal
$S_D(f, \phi_r)$	Doppler power spectral density
$S_M(\tau, \phi_r)$	multi-path intensity profile at time τ and angle ϕ_r
S_ϕ	angular spread
σ_0	interference
τ_e	fixed maximum excess delay
$\sigma_{mai,n}^2$	MAI variance
$\sigma_{si,n}^2$	self interference variance
$\sigma_{ni,n}^2$	AWGN variance
σ_m	GFD fading parameter
σ_T	total interference variance
$\sigma_{angular}^2$	angular spread variance
σ^2	noise variance
σ	noise process standard deviation
σ_s	standard deviation of scatterers
σ_τ	root mean square (rms) delay spread
$s(t)$	received signal power envelope
$s_k(t)$	transmitted signal for user k
t	time
$T(l, i, d)$	state transition matrix elements, as function of number of paths of length l, input weight i, and output weight d
$\bar{\tau}_m$	mean excess delay
τ_0	minimum path delay
$\tau_l^{(k)}(t)$	propagation delay of path l from user k
τ_m	maximum excess delay
τ	delay
$\tau_l^{(k)}$	time delay for specific multipath
θ_k	carrier phase
$tr(\mathbf{R})$	trace of $\mathbf{R}$
T_0	coherence time
T_c	chip period
T_s	symbol period
T_{samp}	sampling period
$T(L, I, D)$	transfer function

$\mathbf{u}^{(n)}$	ideal linear detector output for interval n
$\mathbf{u}$	output of linear detector
U_s	RAKE output
$\mathbf{U}_n$	upper right of $\mathbf{D}_{n,n}$
$\mathbf{U}_{n,n+1}$	upper right matrix for interval n
$\mathbf{U}$	upper right of $\mathbf{R}$
v	mobile speed
V	system load
ν	ratio of R to R_D
$\mathbf{v}^{(n)}$	inner iteration output vector
ϱ_l	relative size of peak
ς_l	angular location of peak l
$\varphi_l^{(k)}(i))$	phase shift for specific multipath
ϖ_l	width of peak l
$\mathbf{w}_q^{(j)}$	weight vector
W	window size
X_σ	zero-mean Gaussian random variable denoting additional p loss
X_{ck}	in-phase Gaussian random variable
X_{sk}	quadri-phase Gaussian random variable
x	encoded bit
$\mathbf{y}^{(n)}$	decision statistic for interval n
$\mathbf{y}(s)$	decision statistic at stage s
$\mathbf{y}$	Rake combined decision statistic
$\mathbf{y}_c$	vector of decision statistic after arbitrary combining
$y_i^{(k)}$	decision statistic for user k symbol interval i
ζ	vector of matched filter outputs
$\zeta_{m_D}^{(j)}(i)$	detector output on m_Dth diversity branch
Z	number of constituent RSC encoders utilized in turbo encoder

Appendix C
Typical Power Delay Profiles

The following four delay power profiles have been proposed in COST-207 as being representative of land mobile propagation environments [45]:

A. *Rural Area*

$$S_M(\tau) = \begin{cases} e^{-9.2\tau} & \text{for } 0 \leq \tau \leq 0.7 \\ 0 & \text{elsewhere} \end{cases} \quad (C.1)$$

Typical Urban Area

$$S_M(\tau) = \begin{cases} e^{-\tau} & \text{for } 0 \leq \tau \leq 7 \\ 0 & \text{elsewhere} \end{cases} \quad (C.2)$$

Bad Urban Area

$$S_M(\tau) = \begin{cases} e^{-\tau} & \text{for } 0 \leq \tau \leq 5 \\ 0.5e^{5-\tau} & \text{for } 5 < \tau \leq 10 \\ 0 & \text{elsewhere} \end{cases} \quad (C.3)$$

Hilly Terrain

$$S_M(\tau) = \begin{cases} e^{-3.5\tau} & \text{for } 0 \leq \tau \leq 2 \\ 0.1e^{15-\tau} & \text{for } 15 \leq \tau \leq 20 \\ 0 & \text{elsewhere} \end{cases} \quad (C.4)$$

In all cases, the delay time variable τ is given in μs.

Appendix D
Correlated Multivariate Gamma Distribution

In mobile communication systems the most frequently used statistical models to describe the amplitude fading process are Rayleigh, Rician and Nakagami distributions. When the *power* of the fading amplitude is of interest, these statistical fading models are all related to the gamma distribution. In diversity based systems the correlated multivariate gamma distribution is of interest. With reference to Figure 5.17, this appendix presents a very general result when arbitrary Nakagami fading, arbitrary correlation and arbitrary signal powers are present on each MRC receive diversity branch.

D.1 CORRELATED MULTIVARIATE GAMMA DISTRIBUTION

Following Figure 5.17, the following assumptions are made in deriving the model:

- arbitrary signal power, Ω_l, on branch l,
- arbitrary correlation, ρ_{kl}, between branches k and l, and
- arbitrary Nakagami fading, m_l, on each branch.

Two well-known correlation models are the constant correlation model, where

$$\rho_{ij} = \rho \qquad \forall \qquad i,j = 1, 2, \cdots, M, \tag{D.1}$$

and the exponential correlation model where

$$\rho_{ij} = \rho^{|i-j|} \qquad \forall \qquad i,j = 1, 2, \cdots, M. \tag{D.2}$$

The constant correlation model may approximate closely spaced diversity antennas. An example of a three-element circular symmetric antenna array that gives rise to a constant correlation matrix is given in [295].

When the diversity signals are taken from a configuration which, in some physical sense, is equi-spaced (either in space, time, frequency, etc), the correlation model can be exponential. The validity of this model stems from the assumption that, given the stationary nature of the overall diversity process (assuming statistical equivalence of the signals), the correlation between a pair of signals decreases as the separation between them increases (see Chapter 4 for a more detailed discussion on this issue).

For a fixed set of received fading amplitudes $\{\beta^{(k)}\}$, the random variables X_{ck} and X_{sk} are normally assumed Gaussian, with received power given by

$$S = \sum_{k=1}^{M} \beta_k^2 = \sum_{k=1}^{M} \{X_{ck}^2 + X_{sk}^2\} = \sum_{k=1}^{M} S_k, \tag{D.3}$$

with S_k the instantaneous power of the kth channel. We note that

$$E\{\beta_k^2\} = E\{S_k\} = \Omega_k. \tag{D.4}$$

In terms of received SNR we can write

$$\gamma_b = \frac{E_b}{N_0} \sum_{k=1}^{M} \beta_k^2 = \sum_{k=1}^{M} \gamma_k, \tag{D.5}$$

as the instantaneous SNR per bit, and

$$\overline{\gamma}_k = \frac{E_b}{N_0} E\{\beta_k^2\} = \frac{E_b}{N_0} \Omega_k, \tag{D.6}$$

the average SNR of the kth diversity branch.

The general characteristic function for an M branch MRC diversity system can be derived as

$$\Phi_S(t) = \Pi_{k=1}^{M} \left| \mathbf{I}_{M_k \times M_k} - it \mathbf{D}_{M_k}\left(\bar{m}^{-1}\right) \mathbf{D}_{M_k}(\Omega_k) \mathbf{J}_{M_k \times M_k} \right|^{-(m_k - m_{k+1})}, \tag{D.7}$$

where $m_M + 1 = 0$, and

$$\begin{aligned} \mathbf{D}_{M_k}(\Omega) &= \text{diag}\{\Omega_1, \Omega_2, \cdots, \Omega_{M_k}\}, \\ \mathbf{D}_{M_k}\{\bar{m}^{-1}\} &= \text{diag}\{m_1^{-1}, m_2^{-1}, \cdots, m_{M_k}^{-1}\}, \end{aligned} \tag{D.8}$$

and $\mathbf{J}_{M_k \times M_k}$ is an $M_k \times M_k$ correlation matrix, given by

$$\begin{aligned} \mathbf{J}_{k,l} &= \rho_{kl} \sqrt{\frac{m_l}{m_k}}, \ k \geq l, \\ k &= 1, 2, \cdots, M. \end{aligned} \tag{D.9}$$

APPENDIX D: CORRELATED MULTIVARIATE GAMMA DISTRIBUTION

The restriction

$$m_1 \geq m_2 \geq \cdots \geq m_M \tag{D.10}$$

applies and the correlation matrix **J** is valid for

$$-\frac{1}{M-1} < \rho < 1. \tag{D.11}$$

The multivariate gamma distribution is finally obtained by taking the inverse Fourier transform

$$p_S(s) = \frac{1}{2\pi} \int_{-\infty}^{\infty} \Phi_S(t) e^{-its} dt. \tag{D.12}$$

It is emphasized again that the characteristic function of (D.7) is very general and valid within the constraints of (D.10) and (D.11).

D.1.1 Example

In this example we show how to calculate the pdf for the sum of $M = 4$ MRC receive diversity signals. We begin by calculating the characteristic function given in (D.7), and then taking the inverse Fourier transform as shown in (D.12).

In our calculations it is important to adhere to the constraints given by (D.10) and (D.11). For our example the fading parameters on the k diversity branches is arbitrarily chosen as $\{m_k\} = 4$ with average received power on each branch $\{\Omega_k\} = 1$. The correlation matrix is given by

$$\mathbf{J} = \begin{pmatrix} 1 & \rho_{21}\sqrt{\frac{m_2}{m_1}} & \rho_{31}\sqrt{\frac{m_3}{m_1}} & \rho_{41}\sqrt{\frac{m_4}{m_1}} \\ \rho_{21}\sqrt{\frac{m_2}{m_1}} & 1 & \rho_{32}\sqrt{\frac{m_3}{m_2}} & \rho_{42}\sqrt{\frac{m_4}{m_2}} \\ \rho_{31}\sqrt{\frac{m_3}{m_1}} & \rho_{32}\sqrt{\frac{m_3}{m_2}} & 1 & \rho_{43}\sqrt{\frac{m_4}{m_3}} \\ \rho_{41}\sqrt{\frac{m_4}{m_1}} & \rho_{42}\sqrt{\frac{m_4}{m_2}} & \rho_{43}\sqrt{\frac{m_4}{m_3}} & 1 \end{pmatrix}. \tag{D.13}$$

In general $\rho_{ij} = \rho_{ji}$ due to symmetry (i.e. $\rho_{12} = \rho_{21}$ etc.). In our example, let $\{\rho_{ij}\} = 0.65$. The correlation matrix **J** therefore reduces to

$$\mathbf{J} = \begin{pmatrix} 1 & 0.65 & 0.65 & 0.65 \\ 0.65 & 1 & 0.65 & 0.65 \\ 0.65 & 0.65 & 1 & 0.65 \\ 0.65 & 0.65 & 0.65 & 1 \end{pmatrix}. \tag{D.14}$$

The calculation of $D_{M_k}(\Omega)$ and $D_{M_k}(\bar{m}^{-1})$ follows trivially from (D.8). By calculating the inverse Fourier transform of (D.7) numerically, the pdf as indicated in Figure D.1 is obtained. Figure D.1 also displays the pdf for different values of ρ.

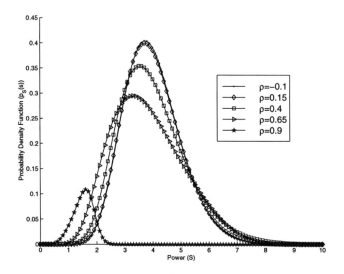

Figure D.1. Probability density function for $M = 4$ and $m = 4$

D.1.2 Special Case

For the constant correlation model given by (D.1), where the received signal strength is constant, i.e. $\{\Omega_k\} = 1$, the fading on the branches is constant, i.e. $\{m_k\} = 1$, and making use of the identity

$$|\mathbf{B}| = [1 - a(1-b)]^{(k-1)} \cdot [1 - a(1 - b + bk)], \qquad (D.15)$$

with

$$\mathbf{B} = \begin{pmatrix} 1-a & -ab & \cdots & -ab \\ -ab & 1-a & \cdots & -ab \\ -ab & -ab & \cdots & -ab \\ -ab & -ab & \cdots & 1-a \end{pmatrix}_{k \times k}, \qquad (D.16)$$

the characteristic function of (D.7) reduces to

$$\Phi_S(t) = [1 - it\Omega^2(1-\rho)]^{-(M-1)} \cdot [1 - it\Omega^2(1 - \rho + \rho M)]^{-1}. \qquad (D.17)$$

Taking the inverse Fourier transform of (D.17), gives a closed form expression for the pdf of interest as

$$p_S(s) = \frac{1}{\Omega^2 \Gamma(M)} \left(\frac{s}{\Omega^2}\right)^{M-1} \qquad (D.18)$$

$$\times \frac{\exp(-\frac{s}{(1-\rho)\Omega^2}) \cdot {}_1F_1(1, M, \frac{\rho M s}{(1-\rho)(1-\rho+\rho M)\Omega^2})}{(1-\rho)^{(M-1)}(1-\rho+\rho M)},$$

where $_1F_1(\cdot)$ is the confluent hyper geometric function [2].

Appendix E
Turbo Code Input-Output CPDF

E.1 INPUT-OUTPUT WEIGHT ENUMERATOR RECURSION

We use the notation and terminology introduced by Divsalar et al. in [55]. The turbo code is the parallel concatenation of the separate rate $R_c = 1$ components, which is referred to as constituent code fragments.

The constituent non-trivial 4-state (2^m) code fragment is completely characterized by its state transition matrix, $\mathbf{A}(L, I, D)$ where

$$\mathbf{A}_{g_{ff}/g_{fb}} = \begin{pmatrix} L & LID & 0 & 0 \\ 0 & 0 & LD & LI \\ LID & L & 0 & 0 \\ 0 & 0 & LI & LD \end{pmatrix}, \tag{E.1}$$

for a given constituent code, denoted by $t(l, i, d)$, the number of paths of length l, input weight i, and output weight d, starting in the all-zero state. Then the corresponding transfer function, or complete path enumerator is defined by [103]

$$\begin{aligned} \mathbf{T}(L, I, D) &= \sum_{l \geq 0} \sum_{i \geq 0} \sum_{d \geq 0} L^l I^i D^d \cdot t(l, i, d) \\ &= \left[(\mathbf{I} - \mathbf{A}(L, I, D))^{-1} \right]. \end{aligned} \tag{E.2}$$

The first element of matrix given in (E.2) produces the transfer function of the constituent code

$$T(L, I, D) = \mathbf{T}^{(1,1)}(L, I, D) = \frac{T_N}{T_D}. \tag{E.3}$$

where

$$T_N = 1 - LD - L^2 I - L^3 (D^2 - I^2), \tag{E.4}$$

and

$$T_D = 1 - L(1+I) - L^3(D^2 - I - I^2 - I^3 D^2)$$
$$- L^4(D^2 - I^2 - I^2 D^4 + I^4 D^2). \quad \text{(E.5)}$$

For a constituent code the path input/output weight index, is used to determine the output weight probability distribution function. For the constituent encoder, the path input/output weight index, is

$$\begin{aligned}
t(l,i,d) &= t(l-1, i-1, d) + t(l-1, i, d) \\
&+ t(l-3, i-3, d-2) - t(l-3, i-2, d) - t(l-3, i-1, d) \\
&+ t(l-3, i, d-2) \\
&- t(l-4, i-4, d-2) + t(l-4, i-2, d-4) + t(l-4, i-2, d) \\
&- t(l-4, i, d-2) \\
&+ \delta(l,i,d) - \delta(l-1, i-1, d) - \delta(l-2, i-1, d) \\
&- \delta(l-3, i, d-2) + \delta(l-3, i-2, d) \quad \text{(E.6)}
\end{aligned}$$

with initial conditions such that $t(l,i,d) = 0$ for any negative index, where $\delta(l,i,d) = 1$ if $l = i = d = 0$ and $\delta(l,i,d) = 0$ otherwise.

E.2 INPUT-OUTPUT CONDITIONAL PROBABILITY DENSITY FUNCTION

In this section the conditional probability of producing a codeword fragment of weight d given a randomly selected input sequence of weight i is evaluated. This is given by

$$p(d \mid i) = \frac{t(N_{tc}, i, d)}{\sum_{d'} t(N_{tc}, i, d')} = \frac{t(N_{tc}, i, d)}{\binom{N_{tc}}{i}}. \quad \text{(E.7)}$$

The CPDF of the constituent recursive convolutional encoder is shown in Figure E.1, for an interleaver size, $N_{tc} = 100$. Since the code fragment only produces even output weights, the uneven weight probabilities are not shown. Similar distributions are obtained when the interleaver size is increased. We refer to the latter as the "unconditioned" CPDF [1]. It was shown in [55] that, given a sufficient input codeword weight i, the CPDF approaches a binomial distribution. Given a balanced source and given that N_{tc} is sufficiently large (typically > 100) this condition will always be met [2]. Thus, in addition to the calculated weight distributions the *conditioned* CPDF, described as a binomial probability distribution with probability $1/2$ (taken over $N_{tc} = 100$ trails) is also shown as reference. In the probability of bit error calculations this *conditioned* CPDF [3] will be assumed in all calculations.

If the interleavers of the encoder are selected randomly and independently, the CPDF $p(d \mid i)$ that any input sequence **u** of weight i will be mapped into

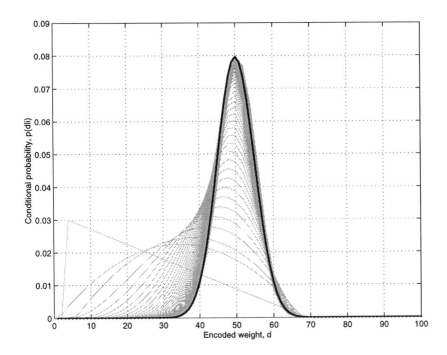

Figure E.1. Conditional probability density function for output weight d of constituent recursive convolutional encoder given input weight i.

code fragments of weights d^s, d^p is

$$R_c = 1/2 : p(d \mid i) = p^s(d_s \mid i) * p^p(d_p \mid i)$$
$$R_c = 1/3 : p(d \mid i) = p^s(d_s \mid i) * p^p(d_p \mid i) * p^p(d_p \mid i), \quad \text{(E.8)}$$

where $*$ denotes convolution.

In (E.7), $p^s(d_s \mid i)$ corresponds to the systematic output weight probability, and $p^p(d_p \mid i)$ corresponds to the non-punctured parity output weight probability. Therefore, the total codeword output weight to the all-zero codeword is $d = d_s + d_p$ and $d = d_s + 2d_p$, for the rate-1/2 and rate-1/3 coders, respectively. The extension to the low-rate code, e.g. $R_c = 1/6$ follows naturally.

In the calculation of the BER for turbo codes of different code rate we have to integrate over the total code input-output weight probability distribution, denoted by $\tilde{p}(d_0, d_i, \ldots, d_Z \mid i)$. This is achieved by the introduction of the conditional expectation $E_{d \mid i}\{\cdot\}$ which should be taken over the probability distribution $\tilde{p}(d_0, d_i, \ldots, d_Z \mid i)$. Here, Z constitutes the number of constituent code fragments.

Notes

1. We refer to the "unconditioned" CPDF, as the Divsalar CPDF.

2. For situations where the input weight may be insufficiently distributed the source data can be scrambled by a Pseudo-Noise sequence with "chip" duration equal to the bit duration.

3. We refer to the "conditioned" CPDF, as the Binomial CPDF.

Appendix F
Turbo Decoding

In this appendix we discuss the concepts of iterative turbo decoding by means of the maximum *a posteriori* (MAP) algorithm. The MAP algorithm calculates the *a posteriori* probabilities, required in the iterative decoding mechanism, directly.

F.1 TURBO DECODING

The problem of estimating the states of a Markov process observed in the presence of noise has two well known trellis-based solutions — the Viterbi algorithm [74] and the (symbol-by-symbol) MAP algorithm [14, 15]. The two algorithms differ in their optimality criterium. The Viterbi algorithm finds the most probable transmitted sequence $\hat{\mathbf{x}}$ given the received sequence $\mathbf{y}$

$$\hat{\mathbf{x}} = \arg\left\{\max_{x} P[\mathbf{x} \mid \mathbf{y}]\right\}. \tag{F.1}$$

The MAP algorithm, on the other hand, attempts to find the most likely transmitted symbol x_i, given the received sequence $\mathbf{y}$

$$\hat{x}_i = arg\left\{\max_{x_i} P[x_i \mid \mathbf{y}]\right\}. \tag{F.2}$$

The problem of decoding turbo codes involves the joint estimation of two (for the punctured rate-1/2 case) or more Markov processes, one for each constituent code. While in theory it is possible to model a turbo code as a single Markov process, such a representation is extremely complex and does not lend itself to computationally tractable decoding algorithms. Thus, turbo decoding first independently estimates the individual Markov processes using two trellis based decoding algorithms. Because the two Markov processes are linked by an interleaver, additional gain can be achieved by sharing information between the two decoders in an iterative fashion. More specifically, the output of one decoder can be used as *a priori* information by the other decoder. If the outputs of the individual decoders are in the form of hard decisions, then there is little advantage in sharing information. However, if soft decisions are produced by the individual decoders, considerable gain can be achieved by performing multiple iterations of decoding.

F.1.1 Extrinsic and Intrinsic Information

F.1.1.1 Extrinsic Information. Extrinsic information in relation to a data symbol, is information from other symbols in the codeword sequence imposed by the code constraints without using any information concerning the symbol itself. In other words, the extrinsic information relating to a specific symbol is only determined by its surrounding symbols.

The term soft output refer to internal variables of the decoder, indicating a measure of decoding reliability of bits instead of providing hard decisions.

F.1.1.2 Intrinsic Information. Intrinsic information in relation to a data symbol is *a priori* information of the symbol without any code constraints. This information is used by the constituent decoders as additional information related to each code symbol. In iterative decoding, the extrinsic information provided by the previous decoding step becomes the *a priori* information of the current decoding stage.

F.1.2 Soft-Input/Soft-Output (SISO) Decoding Algorithm

Soft decisions typically take the form of (*a posteriori*) log-likelihood ratios (LLRs), Λ of the form

$$\Lambda_i = \ln \frac{P[u_i = 1 \mid \mathbf{y}]}{P[u_i = 0 \mid \mathbf{y}]}, \tag{F.3}$$

where u_i is the information input bit.

A decoding algorithm that accepts *a priori* information at its input and produces *a posteriori* information at its output is called a soft-input/soft-output (SISO) decoding algorithm.

As shown in Figure F.1, a SISO decoder for a rate-1/2 RSC code accepts three inputs — the systematic observations $y_i^{(s)}$ (agreeing with transmitted bit $x_i^{(1)}$), the parity observations $y_i^{(p)}$ (agreeing with either transmitted bits $x_i^{(2)}$ or $x_i^{(3)}$, or both), and the *a priori* information z_i which is derived from the other decoder's output. The decoder then produces an output in the LLR form.

The log-likelihood at the output of a SISO decoder using this channel model can be factored into three terms [100]:

$$\Lambda_i = \frac{4 \, a_i^{(s)} \, E_s}{N_0} y_i^{(s)} + Z_i + L_i \tag{F.4}$$

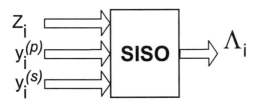

Figure F.1. Soft-Input/Soft-Output (SISO) decoder for a rate-1/2 RSC code.

where the term L_i is called the extrinsic information, and $a^{(s)} = a^{(1)}$ is the fading amplitude associated with the received systematic observation, $y_i^{(s)}$. While the first two terms in (F.4) are related to the systematic channel observation $y_i^{(s)}$ and information derived by the other decoder Z_i, the extrinsic information represents new information derived by the current stage of decoding. In order to prevent positive feedback, it is important that only the extrinsic information is passed from one decoder to the other. Thus, the *a priori* information at the input of one encoder is found by subtracting two values from its output — a value proportional to the encoder's systematic input as well as the other encoder's *a priori* input

$$L_i = \Lambda_i - \frac{4 a_i^{(s)} E_s}{N_0} y_i^{(s)} - Z_i. \qquad (\text{F.5})$$

The schematic for a standard turbo decoder is shown in Figure F.2. The first decoder receives the systematic channel observation $\mathbf{y}^{(s)} = \{y_1^{(1)}, y_2^{(1)}, \ldots, y_{N_{tc}}^{(1)}\}$, observations of the first encoder's parity bits $\mathbf{y}^{(p)} = \{y_1^{(2)}, y_2^{(2)}, \ldots, y_{N_{tc}}^{(2)}\}$, and *a priori* information $\mathbf{Z}^{(1)}$ derived from the second decoder's output. The first decoder produces the LLR, $\Lambda^{(1)} = \{\Lambda_1^{(1)}, \Lambda_2^{(1)}, \ldots, \Lambda_{N_{tc}}^{(1)}\}$.

The extrinsic information of the first decoder $\mathbf{L}^{(1)}$ is found by subtracting the weighted systematic and *a priori* inputs from the first decoder's output. The extrinsic information is interleaved, and used as *a priori* information by the second decoder. The second decoder also receives the interleaved and weighted systematic channel observation $\bar{\mathbf{r}}^{(1)}$ and weighted observations of the second encoder's parity bits $\mathbf{r}^{(3)}$.

The second decoder produces the LLR, $\Lambda^{(2)}$, from which the second decoder's weighted systematic and *a priori* inputs are subtracted to produce the extrinsic information $\mathbf{L}^{(2)}$. The extrinsic information produced by the second decoder is de-interleaved and used as the *a priori* input to the first decoder during the next iteration. After I_{tc} iterations, the final estimate of the message is found

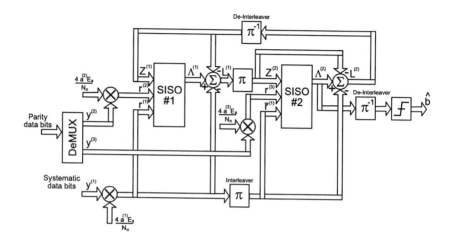

Figure F.2. Turbo decoder block diagram.

by de-interleaving and hard-limiting the output of the second decoder

$$\hat{u}_i = \begin{cases} 1 & \text{if } \Lambda^{(2)}_{\pi^{-1}(i)} \geq 0 \\ 0 & \text{if } \Lambda^{(2)}_{\pi^{-1}(i)} < 0 \end{cases} . \tag{F.6}$$

F.2 ITERATIVE DECODING ALGORITHMS

Several different algorithms can be used to implement the SISO decoder components of a turbo decoder. These algorithms can be partitioned into main two classes, depending on whether they were derived from the Viterbi algorithm or the MAP algorithm. In general, the algorithms based on the MAP algorithm are more complex and perform better than the algorithms based on the Viterbi algorithm.

Of recent interest in this area is the use of general graphical models (e.g. Tanner graphs) to describe concatenated codes [136, 319]. Just as trellis-based code descriptions are naturally matched to Viterbi decoding, code descriptions based on Tanner graphs (which may be viewed as generalized trellises) are naturally matched to iterative decoding.

Here, the discussion is restricted to the MAP algorithm. Specifically, the MAP decoder proposed by Robertson [224, 225] is considered in detail.

F.2.1 Maximum A Posteriori (MAP) Algorithm

The symbol-by-symbol MAP Algorithm was formally presented in 1974 by Bahl *et al.* as an alternative to the Viterbi algorithm for decoding convolutional codes

[14, 15]. While the Viterbi algorithm minimizes the probability of sequence error, the MAP algorithm minimizes the probability of symbol error. A different version of the MAP algorithm was presented in 1970 by Abend and Ritchman for the same application [1]. The algorithm in [14, 15] requires a forward and backward recursion and is therefore suitable for block-oriented processing. The algorithm of [1] only requires a forward recursion and is suitable for continuous processing, although it is more complex and generally requires more total storage than the forward-backward version of the algorithm. Following the convention of [310], we will denote the block-oriented MAP algorithm as the type-I MAP algorithm and the continuous MAP algorithm as the type-II MAP algorithm. Because the type-II MAP algorithm is more complex and requires more storage than the type-I MAP algorithm, and because turbo encoding is a block-oriented process, we treat only the type-I MAP algorithm. Thus, unless otherwise specified, the term "MAP algorithm" refers to the type-I MAP algorithm.

The goal of the MAP algorithm is to first find the *a posteriori* probability of each state transition, message bit, or code symbol produced by the underlying Markov process, given the noisy observation $\mathbf{y}$. Once the *a posteriori* probabilities are calculated for all possible values of the desired quantity, a hard decision is made by taking the quantity with highest probability. When used for turbo decoding, the MAP algorithm calculates the a posteriori probabilities of the message bits, $P[u_i = 1 \mid \mathbf{y}]$ and $P[u_i = 0 \mid \mathbf{y}]$, which are then written as LLRs according to (F.3). The version of the MAP algorithm used for turbo decoding does not make hard-decisions on the message bits until after the last decoder iteration.

F.2.1.1 Calculation of Branch-Metric, γ.

Before finding the *a posteriori* probabilities for the message bits, the MAP algorithm first finds the probability $P[s_i \to s_{i+1} \mid \mathbf{y}]$ of each valid state transition given the noisy channel observation $\mathbf{y}$. From the definition of conditional probability

$$P[s_i \to s_{i+1} \mid \mathbf{y}] = \frac{P[s_i \to s_{i+1}, \mathbf{y}]}{P[\mathbf{y}]}. \quad (F.7)$$

The properties of the Markov process can be used to partition the numerator as

$$P[s_i \to s_{i+1}, \mathbf{y}] = \alpha(s_i)\, \gamma(s_i \to s_{i+1})\, \beta(s_{i+1}), \quad (F.8)$$

where

$$\alpha(s_i) = P[s_i, (y_0, \ldots, y_{i-1})] \quad (F.9)$$
$$\gamma(s_i \to s_{i+1}) = P[s_{i+1}, y_i \mid s_i]$$
$$\beta(s_{i+1}) = P[(y_{i+1}, \ldots, y_{L-1}) \mid s_{i+1}].$$

From (F.9), $\gamma(s_i \to s_{i+1})$ is the branch metric associated with the transition $s_i \to s_{i+1}$, which can be expressed as

252   SPACE-TIME PROCESSING FOR CDMA

$$\begin{aligned}\gamma(s_i \to s_{i+1}) &= P[s_{i+1} \mid s_i] \, P[y_i \mid s_i \to s_{i+1}] \\ &= P[b_i] \, P[y_i \mid x_i],\end{aligned} \qquad \text{(F.10)}$$

where b_i and x_i are the message and output (respectively) associated with the state transition $s_i \to s_{i+1}$. Note that if the states s_i and s_{i+l} are not connected in the trellis diagram, then the above probability is zero.

Figure F.3 depicts a graphical illustration of the computation of the branch metric, γ. The latter is also known as the branch transitions probabilities.

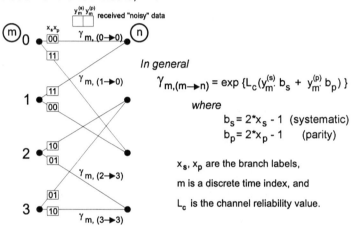

Figure F.3.   Graphical illustration of the computation of the branch metric, γ.

$P[b_i]$ of (F.10) is obtained using the *a priori* information Z_i and, $P[y_i \mid x_i]$ is a function of the modulation and channel model.

F.2.1.2   Calculation of Forward Recursion, α and Backward Recursion, β.

The probability $\alpha(s_i)$ can be found using the forward recursion

$$\alpha(s_i) = \sum_{s_{i-1}} \alpha(s_{i-1}) \, \gamma(s_{i-1} \to s_i). \qquad \text{(F.11)}$$

Likewise, $\beta(s_i)$ can be found with the backward recursion

$$\beta(s_i) = \sum_{s_{i+1}} \beta(s_{i+1}) \, \gamma(s_i \to s_{i+1}). \qquad \text{(F.12)}$$

Figures F.4 and F.5 depict graphical illustrations of the computation of, respectively, the forward recursion, α, and the backward recursion, β.

APPENDIX F: TURBO DECODING   253

Calculation of forward recursion, α :

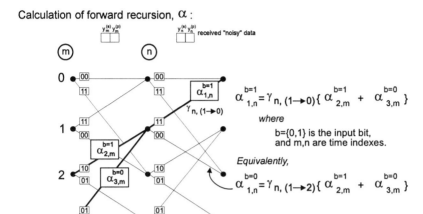

$$\alpha_{1,n}^{b=1} = \gamma_{n,(1\to 0)} \{ \alpha_{2,m}^{b=1} + \alpha_{3,m}^{b=0} \}$$

where

b={0,1} is the input bit,
and m,n are time indexes.

Equivalently,

$$\alpha_{1,n}^{b=0} = \gamma_{n,(1\to 2)} \{ \alpha_{2,m}^{b=1} + \alpha_{3,m}^{b=0} \}$$

Figure F.4. Graphical illustration of the computation of the forward recursion, α.

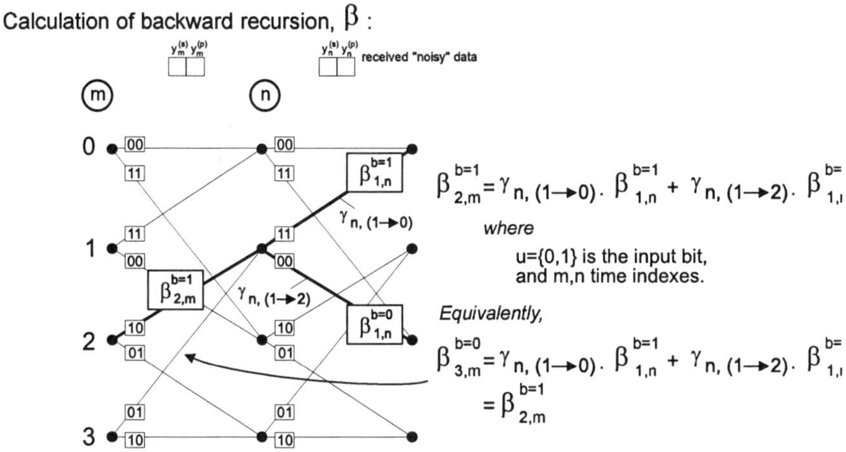

Calculation of backward recursion, β :

$$\beta_{2,m}^{b=1} = \gamma_{n,(1\to 0)} \cdot \beta_{1,n}^{b=1} + \gamma_{n,(1\to 2)} \cdot \beta_{1,1}^{b=}$$

where

u={0,1} is the input bit,
and m,n time indexes.

Equivalently,

$$\beta_{3,m}^{b=0} = \gamma_{n,(1\to 0)} \cdot \beta_{1,n}^{b=1} + \gamma_{n,(1\to 2)} \cdot \beta_{1,1}^{b=}$$
$$= \beta_{2,m}^{b=1}$$

Figure F.5. Graphical illustration of the computation of the backward recursion, β.

F.2.1.3 LLR Calculation, Λ. Once the *a posteriori* probability of each state transition $P[s_i \to s_{i+1} \mid \mathbf{y}]$ is found, the message bit probabilities can be determined as

$$P[b_i = 1 \mid \mathbf{y}] = \sum_{S_1} P[s_i \to s_{i+1} \mid \mathbf{y}], \qquad (F.13)$$

and

$$P[b_i = 0 \mid \mathbf{y}] = \sum_{S_0} P[s_i \to s_{i+1} \mid \mathbf{y}], \qquad (F.14)$$

where $S_1 = \{s_i \to s_{i+1} : b_i = 1\}$ is the set of all state transitions associated with a message bit of 1, and $S_0 = \{s_i \to s_{i+1} : b_i = 0\}$ is the set of all state transitions associated with a message bit of 0. The LLR then becomes

$$\Lambda_i = \ln \frac{\sum_{S_1} \alpha(s_i)\, \gamma(s_i \to s_{i+1})\, \beta(s_{i+1})}{\sum_{S_0} \alpha(s_i)\, \gamma(s_i \to s_{i+1})\, \beta(s_{i+1})}. \qquad (F.15)$$

Note that the term $P[y]$ in (F.7) disappears when the ratio is taken, and therefore it is not explicitly calculated.

F.2.1.4 Tail Termination. As mentioned earlier tail (trellis) termination is a problem in turbo coding. This is because additional bits are required to force the code to a known state at the end of a decoding block. The initial state of the block is used to initialize α_0 for all states, of the forward recursion. The known final state of the block is used to initialize $\beta_{N_{tc}}$ for all states, in the backward recursion.

In the WCDMA Simulation Environment (see Appendix H) a termination scheme which removes transmission of the tail of both encoders is used. This is done simply by setting the initial value of the backward recursion to a constant.

To summarize, the MAP algorithm operates as follows

1. Forward recursion:

   (a) Initialize $\alpha(s_i)$ according to

   $$\alpha(s_0) = \begin{cases} 1 & \text{if } s_0 = 0 \\ 0 & \text{if } s_0 \neq 0 \end{cases}. \qquad (F.16)$$

   (b) Let $i = 1$.
   (c) Determine $\alpha(s_i)$ using (F.11).
   (d) Increment i. If $i < L$ return to Step 1.(c). Otherwise, continue to Step 2.

2. Backward recursion:

   (a) If the trellis is terminated, initialize $\beta(s_i)$ according to

   $$\beta(s_L) = \begin{cases} 1 & \text{if } s_L = 0 \\ 0 & \text{if } s_L \neq 0 \end{cases}. \qquad (F.17)$$

Otherwise, if the trellis is not terminated, initialize $\beta(s_i)$ according to

$$\beta(s_L) = \frac{1}{2^{M_{cc}}} \quad \forall s_L. \tag{F.18}$$

(b) Let $i = L - 1$.
(c) Determine $\beta(s_i)$ using (F.12).
(d) Decrement i. If $i > 0$ return to Step 2.(c). Otherwise, continue to Step 3.

3. For $i = (0, \ldots, L-1)$, determine the LLR according to (F.15).

Matlab files that implement the MAP algorithm described above is included with this book. The files are

Appendix G
WCDMA Simulation Environment: Physical Layer

In this section the physical layer for the UMTS-like WCDMA Simulation Environment (see Appendix H) is described. For more information on the ETSI UMTS standard, the interested reader is referred to [65, 68, 67].

The key technical parameters for the WCDMA radio interface are listed in Table G.1. The multiple access scheme is based on wide band direct-sequence CDMA, with a basic chip rate of 4.096 Mcps. The chip rate can be expanded to 8.192 and 16.384 Mcps in order to extend the bit rate above 2 Mbps.

Multiple access scheme	Wideband DS-CDMA
Duplex scheme	FDD
Chip rate	4.096 Mcps (8.192/16.348 Mcps)
Carrier spacing	4.4-5.0 MHz (200 kHz raster)

Table G.1. Key parameters for UMTS.

The physical and transport channels used are divided into two classes, dedicated channels and common channels. A dedicated channel refers to a channel used for point-to-point communications (for example, base station to a single mobile). A common channel refers to a channel used for either point-to-multipoint (for example, base station to all mobiles in cell), or multi-point-to-point communications (for example, many mobiles sharing an uplink channel to the base station). The dedicated physical channels are:

- The dedicated physical data channel (DPDCH) used to carry dedicated data generated at Layer 2 and above.

- The dedicated physical control channel (DPCCH) used to carry Layer 1 control information.

Each connection is allocated one DPCCH and zero, one or several DPDCH's. In addition, the following common physical channels are defined:

- The primary and secondary common control physical channels (CCPCH) are used to carry downlink common channels.
- The synchronization channel (SCH) is used for cell search, and
- The physical random access channel (PRACH).

The following two sections provide a summary of the WCDMA specification for the physical layer of the uplink and downlink.

G.1 UPLINK DESCRIPTION SUMMARY

The uplink DPDCH and DPCCH channels are I/Q multiplexed as shown in Figure G.1. That is, the DPDCH channel is transmitted on the inphase component of the transmitted signal, and the DPCCH channel is transmitted on the quadrature component. The DPDCH carries Layer 2 data, while the DPCCH carries pilot bits, transmit power-control (TPC) commands, and an optional transport-format indicator (TFI) [1]. For each connection between a mobile and the base station, several (or no) DPDCH channels can be in use. However, a single DPCCH is always used. If more than one DPDCH channel is needed by a particular user, multi-code transmission is required. Two DPDCH channels can then be I/Q multiplexed together. Thus to receive a DPDCH channel requires knowledge of its carrier frequency, spreading codes and relative phase (I/Q). The DPCCH carries the control information generated by the physical layer only. Control information generated at higher layers is transferred using a DPDCH channel. Thus the DPCCH carries known pilot symbols for use by the base station receiver, power control information and signal format information.

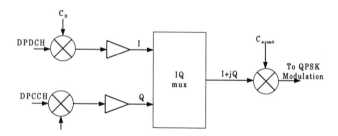

Figure G.1. Uplink spreading and modulation for DPDCH and DPCCH.

As shown in Figure G.2, each frame of length 10 ms is divided into 16 slots of length 0.625 ms, each corresponding to one power-control period (hence the power-control frequency is 1600 Hz).

APPENDIX G: WCDMA SIMULATION ENVIRONMENT: PHYSICAL LAYER   259

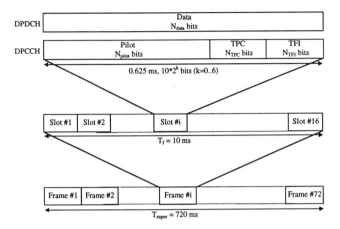

Figure G.2.   Frame structure for the uplink DPDCH/DPCCH channels.

The DPDCH and DPCCH channels are each multiplied by different binary-valued channelisation codes. Each channelisation code has the appropriate length to spread the DPDCH or DPCCH channel to the chip rate. DPDCH channels on the same branch (I/Q) must use different channelisation codes; however, when multi-code transmission is used, DPDCH channels on differing branches (I/Q) can use the same channelisation code. The channelisation codes are orthogonal variable spreading factor (OVSF) codes generated by the code-generation tree shown in Figure G.3.

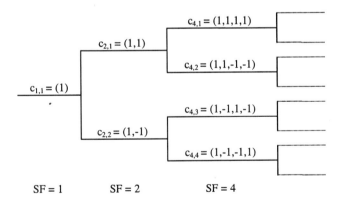

Figure G.3.   Code generation tree for OVSF codes.

The PRACH is the physical channel type used to carry the RACH transport channel. A Slotted ALOHA scheme is used in which the mobile can begin to transmit a packet to the base station at one of 8 time instants within a 10 ms frame. Thus the mobile can begin packet transmissions at time instants separated by 1.25 ms, and aligned with the 10 ms frame structure. A packet consists of a pre-amble, followed by an idle period, followed by a 10 ms message. The entire packet transmission takes 11.25 ms.

At each node of the tree, the upper branch to the right of the node is labeled with the sequence on the branch to the left of the node, followed by the same sequence repeated. The lower branch to the right of the node is labeled with the sequence on the branch to the left of the node, followed by its complement. The code associated with a particular branch of the tree is orthogonal to all the codes in the tree except for the following:

- codes on the path from the specified code to the root of the tree; and

- codes in the sub-tree below the specified code

Thus orthogonal sequences of different lengths (and hence spreading factor) can be generated. The user may thus have several dedicated channels to transmit to the base station. These channels may be of different bit rates. Each channel is assigned an orthogonal channelisation code of the appropriate spreading factor to spread the channel to the chip rate. The chip rate is always 4.096 Mcps for all users and physical channel types. The symbol rate for a particular user's physical channel is thus controlled by the spreading factor used. The spreading factors can vary between 4-256, with a spreading factor of $256/2^k, k = 0, 1, \cdots 6$, carrying 10×2^k bits per slot each. The DPDCH and DPCCH channels are assigned different channelisation codes since they may be of different rates. To control the amount of overhead, the relative power between the DPCCH and DPDCH can be varied. Typical values for the relative power difference are 3 and 10 dB for speech and 384 kbps data respectively. There is no requirement to co-ordinate the allocation of channelisation codes between mobile stations. This is because a second stage of signal spreading using a scrambling code is employed.

The I/Q components of the transmitted signal, after being spread to the chip rate by the channelisation code, are viewed as a single complex-valued signal. A complex multiplication is performed between this signal and a complex valued scrambling code. The UMTS specification, on which the WCDMA simulator is based, allows for the possibility of using either short or long codes in the uplink. The uplink scrambling code can be either long or short. The short scrambling code is a complex code built of two 256-chip long extended codes from the very large Kasami set of length 255. The long scrambling code is a 40 960-chip segment of a Gold code of length $2^{41} - 1$. These sequences are both binary valued. Short scrambling codes are intended for use in cells that support multiuser detector in the base station.

The IQ multiplexing of control and data is used to ensure that EMC problems are minimized in the UE. To minimize interference and maximize capacity, during speech silent periods no data is transmitted. However, pilot bits and power-control commands are needed to keep the link synchronized and power controlled. The IQ multiplexing also avoids pulsing the power with a given frequency. If time multiplexing of control and data were used instead, a 16–Hz tone would be emitted during silent periods.

G.2 DOWNLINK DESCRIPTION SUMMARY

As shown in Figure G.4, the DPDCH and DPCCH are time multiplexed within the downlink radio frames. As in the uplink, the downlink DPDCH contains Layer 2 data, while the DPCCH carries pilot bits, TPC commands and optional TFI. Like the uplink, each frame of length 10 ms is divided into 16 slots of length 0.625 ms, each corresponding to one power-control period. Within each slot, the DPCCH and DPDCH are time multiplexed and transmitted with the same code on both the I/Q branches. The spreading factor for the DPDCH and DPCCH can vary between 4-256, with a spreading factor of $256/2^k, k = 0, 1, \cdots, 6$, carrying a total of 20×2^k bits per slot.

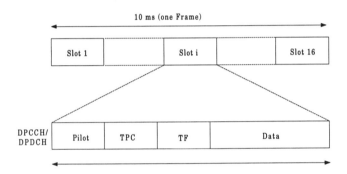

Figure G.4. Frame structure for the downlink DPDCH/DPCCH channels.

The spreading and modulation of the downlink dedicated physical channels are shown in Figure G.5. The DPCCH/DPDCH bits are mapped in pairs to the I and Q branches and spreading to the chip rate is done with the same channelisation code on both I and Q branches. Subsequently scrambling is performed before QPSK modulating the complex signal. Root-raised cosine pulse shaping with a roll-off factor of 0.22 in the frequency domain is used. Channelisation is done using the same type of OVSF codes as for the uplink dedicated physical channels, and the set of codes used can be changed by the network during a connection. The downlink scrambling code is a 40 960 chip segment of a Gold code of length $2^{18} - 1$. There are 512 different segments used for downlink scrambling. These are divided into 16 groups of 32 codes each in order to simplify the cell-search procedure. Each cell is assigned a specific downlink scrambling code at initial deployment. For multi-code transmission, each additional DPCCH and DPDCH is spread and scrambled in a similar way, using a channelisation code that keeps the physical channels orthogonal.

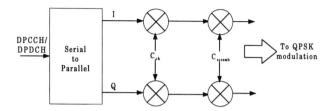

Figure G.5. Downlink spreading and modulation for DPDCH and DPCCH.

In contrast to the uplink, time multiplexing of control and data does not lead to EMC problems in the downlink. Taking into account the fact that all users share the channelisation codes in the downlink, the IQ multiplexing scheme where a whole code is needed for the DPCCH only will use unnecessarily many codes. Hence, time multiplexing is a logical choice in the downlink.

Notes

1. The TFI informs the receiver side what transport-format is used in the current data frame in order to simplify detection, decoding, demultiplexing, etc.

Appendix H
WCDMA Simulation Environment

This appendix provides a brief description of the WCDMA simulation platform which is based on a UMTS-like standard as described in Appendix G. The simulation includes all aspects described in the book applicable to a space-time based CDMA system. Combinations of space-time techniques, such as transmit diversity, receive diversity, RAKE combining, convolutional coding, turbo coding and power control, can be simulated in an easy to use graphical user interface (GUI). The simulation has been written and tested in MATLAB 5.2, running on a WINDOWS 95/98 platform. The complete MATLAB simulation environment is available on floppy disk included with the book, as well as from the Mathworks at

    web: ftp://ftp.mathworks.com/pub/books/vanrooyen

  or by anonymous FTP

    Unix login: ftp ftp.mathworks.com
                     Name: anonymous
                     Guest login ok, send your complete e-mail address
                     Password: (type in e-mail address)
                     cd/pub/books/vanrooyen

H.1   LINK LEVEL SIMULATION

The simulation software is capable of simulating both the uplink and downlink of a UMTS-like WCDMA systems. Figures H.1 and H.2 illustrate the uplink and downlink block diagrams respectively.

As described in Appendix G, a transmission frame consists of multiple slots. In the simulation, the key modules rely on frame and slot based processing for

264 SPACE-TIME PROCESSING FOR CDMA

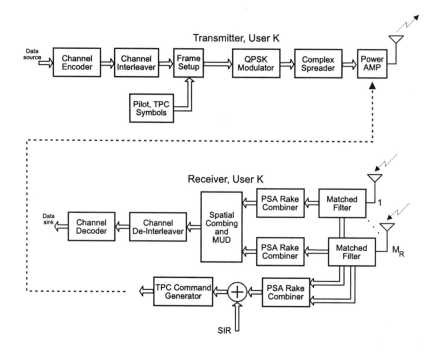

Figure H.1. Overall block diagram of the uplink.

both the receive and transmit functions. In the transmitter, the frame based processing consists of frame encoding and interleaving.

For the uplink a single transmit antenna is used, while the downlink may include M_T transmit antennas, using either CDTD or TDTD. The in-phase and quadrature components of the transmitted signal are multiplied by a random segment of a pre-generated fading channel complex envelope. The channel models implemented are UMTS indoor, outdoor-to-indoor/pedestrian and vehicular models. The resulting signals are then added and finally AWGN is added.

For each user the physical channel data consist of an information sequence with control information. The length of the information sequence and the encoding rate sets the number of binary symbols to be transmitted on the I and Q branches of the modulator. This in turn sets the processing gain of each user. Users with higher information rates will have correspondingly lower spreading gains. In addition to specifying the processing gain, frame interleaver size of the convolutional- and turbo encoding and decoding are also determined by the information data rate.

In the receiver, the received signal is first processed by a chip matched filter. Thereafter, a RAKE receiver consisting of a number of correlators (or fingers),

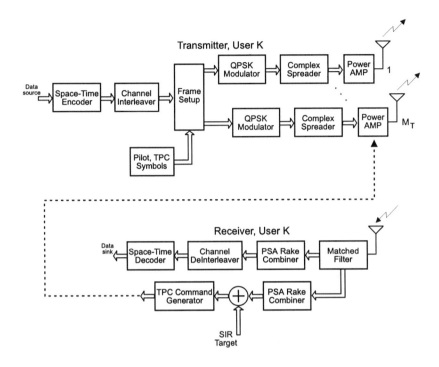

Figure H.2. Overall block diagram of the downlink.

operating in parallel is used to process the received signal. Each finger correlates a shifted version of the received signal with the spreading sequence for the user of interest. The different shifts correspond to the different excess delays for each multi-path component received by the mobile terminal. The outputs of the RAKE fingers must be combined (once per symbol period) to obtain an estimate of the received symbol. For transmit diversity, in addition to the standard operation of the RAKE receiver, channel estimation is performed on each resolved path, and used in a pilot symbol assisted (PSA) RAKE combiner to resolve each of the transmitted streams from the multiple transmit antennas.

Closed loop power control is used on the dedicated channels to reduce the imbalance in received power (near-far effect). Ideally the base station adjusts the transmit power of mobiles such that the base station observes a prescribed SNR. Both pilot and data symbols are used in measuring the instantaneous received signal power, with pilot symbols being used in the measurement of instantaneous interference plus background noise power. The measured SIR is then compared to a target value to generate the transmit power control (TPC) command. This command is sent to the transmitter at the mobile to raise or lower by 1 or 2 dB the transmit power at the end of every slot.

H.1.1 Monte-Carlo Simulation Technique

To enable statistically valid simulation results in reasonable simulation times, Monte Carlo methods are used.

For a given number of users, channel model and link configuration the aim of the simulation is to produce a bit error rate curve as a function of SNR. For each SNR value, the uplink or downlink is simulated until a reliable estimate of the bit error rate at the output of the detector is obtained.

A simulation "loop" is defined as the transmission and reception of a 10 ms frame. Simulation loops are continued until both of the following conditions are satisfied:

- The number of bit errors detected by the receiver is greater than a specified minimum number of errors, and

- The number of simulation loops performed is greater than a specified minimum number of loops.

With the above two conditions met, the simulation will continue until one of the following conditions is true:

- The number of simulation loops reaches a specified maximum number of loops, and

- The current bit error rate is less than a specified minimum rate.

H.1.2 General Simulation Assumptions

- Inter-cell interference is not modeled.

- Narrowband interference is not modeled as a component of the channel model.

- Linear power amplifiers at both the transmitter and receiver.

- The uplink and downlink simulation operates in FDD mode.

- For all receiver types, it is assumed that the receiver can synchronize to the received signal. No synchronization errors are taken into account by the simulation.

- The RAKE receiver is provided with the excess delays for each multi-path component processed. Thus, the simulation does not perform MPC delay estimation.

- The IC based receivers are provided with ideal channel estimates. It is true that the performance of the channel estimator in an IC environment will be better than other cases since the multiuser interference is greatly reduced in the signal supplied to the channel estimator.

- The AS-TDTD transmitter and turbo MAP decoder are provided with estimated channel conditions.

H.1.3 Simulation Cases

Three types of users, each having different service requirements, may be considered. These are indicated in Table H.1.

Parameter	Class 1	Class 2	Class 3
Physical channel rate (uncode)	48 kbps	256 kbps	1024 kbps
Spreading factor, N	32	16	4
FEC rate	1 or 1/3	1 or 1/3	1 or 1/3
Frame Interleaving	10 ms	10 ms	10 ms
DPCCH/DPDCH power	0 dB	0 dB	0 dB

Table H.1. Simulation service classes.

Table H.2 provides a summary of the receivers, transmit diversity, and coding techniques.

Acronym	Description
SICL	Iterated SIC, no clip/linear (3 Iterations)
SICH	Iterated SIC, hard (3 Iterations)
SICL	Iterated SIC, clip (3 Iterations)
PICL	Iterated PIC, no clip/linear (3 Iterations)
PICH	Iterated PIC, hard (3 Iterations)
PICL	Iterated PIC, clip (3 Iterations)
NLMS	Normalized LMS ($\mu = 0.02$)
EMF	Estimated matched filter ($\mu = 0.02$)
NO TD	No transmit diversity
O-CDTD	Orthogonal code division transmit diversity
RR-TDTD	Round-robin time division transmit diversity
AS-TDTD	Antenna selection time division transmit diversity
UNC	Uncoded transmission
RC	Repetition coding
CC	Convolutional coding (128 or 256 states)
TC	Turbo coding (4, 8 or 16 states) MAP decoding, 8 iterations

Table H.2. Summary of simulation features.

H.2 MATLAB SIMULATION SOFTWARE

H.2.1 Getting Started

To run the MATLAB simulation platform, the following steps should be followed:

Step 1 Create a suitable working directory to which the software will be copied. For example: 'c:\wcdmasim'.

Step 2 Copy the 'p-code' files ('*.p') to the working directory.

Step 3 Create the simulation data directory to which 'error' and 'log_file' results will be stored. This directory should be created on the 'C' drive as follows: 'c:\data'.

Step 4 Start MATLAB, and add the directory created under Step 1 to the MATLAB path.

Step 5 Type 'wcdmasim' at the MATLAB command line.

H.2.2 Main Simulation Window

By invoking 'wcdmasim' at the MATLAB command line, the main GUI from which different simulation engines are called from will be opened. A screen capture of this GUI window is depicted in Figure H.3.

H.2.3 Simulation Environment Configuration

By selecting 'Transceiver/Channel Setup', the configuration window, shown in Figure H.4 will be displayed.

The WCDMA transceiver and environment parameters controlled by the GUI are:

- General transceiver parameters:
  - Simultaneous users, K.
  - Users load in a mixed throughput environment, given as percentage of number of simultaneous transmitting users.
  - SNR range and step increments.

- Channel environment parameters:
  - Type: AWGN, UMTS Indoor, UMTS Outdoor-to-Indoor and Pedestrian, and UMTS Vehicular.
  - Average speed and log-normal shadowing variance.

- Monte-Carlo simulation parameters:
  - Minimum number of bit errors.

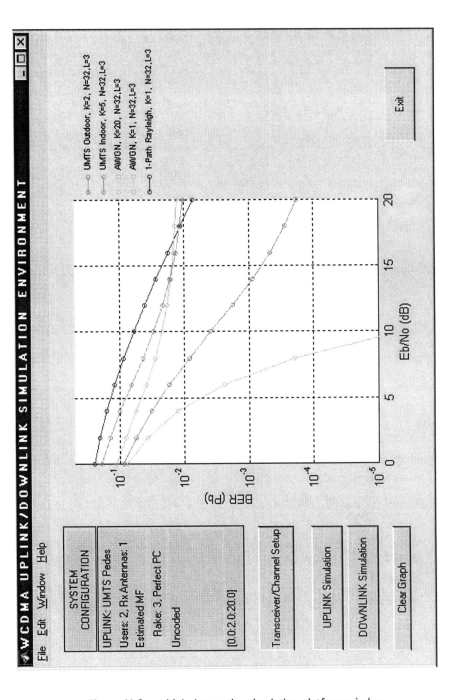

Figure H.3. Main interactive simulation platform window.

270   SPACE-TIME PROCESSING FOR CDMA

- Minimum and maximum number of received frames.

- Parameters common to uplink and downlink:
  - RAKE fingers.
  - Power control algorithm selection.

- Parameters specific to uplink:
  - MRC receiving antennas.
  - Receiver (single- or multiuser detectors):
    * Iterated SIC, No clip.
    * Iterated SIC, Clip.
    * Iterated SIC, Hard.
    * Iterated PIC, No clip.
    * Iterated PIC, Clip.
    * Iterated PIC, Hard.
    * Estimated Matched Filter (EMF).
    * Normalized LMS (NLMS).
  - FEC technique:
    * No coding.
    * Convolutional encoder with soft-input Viterbi decoder.
    * Turbo encoder with iterative MAP decoder (8 Iterations).

- Parameters specific to downlink:
  - Transmitting antennas.
  - Transmit diversity selection:
    * No transmit diversity.
    * O-CDTD.
    * RR-TDTD.
    * AS-TDTD.
  - Receiver:
    * Estimated matched filter (EMF).
    * Normalized LMS (NLMS).
  - FEC technique:
    * No coding.
    * Convolutional encoder with soft-input Viterbi decoder.
    * Turbo encoder with iterative MAP decoder (8 Iterations).

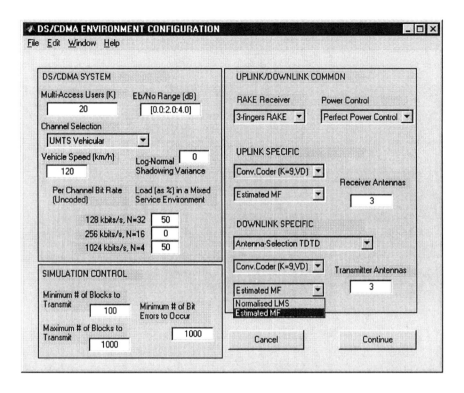

Figure H.4. Simulation platform configuration window.

H.2.4 Example

The following example can be used as a reference describing the operation of the simulation.

Step 1 Type 'wcdmasim' at the MATLAB command line. This will bring up the main interface window, shown in Figure H.3.

Step 2 Click on the 'Transceiver/Channel Setup' button. This will open the configuration window, as shown in Figure H.4.

Step 3 Select the number of users, K.

Step 4 Change the 'E_b/N_0 range in dB' entry to the desired range. Note that this parameter is entered in typical MATLAB style for vectors which is *start:step:end* with *step* defaulting to 1.0 if not specified.

Step 5 Select the desired channel environment.

Step 6 Select the vehicle speed and log-normal shadowing variance.

Step 7 Set up the users load as a percentage. Upon exit the entries will be normalized to a total load of 100 %.

Step 8 Change the simulation control parameters.

Step 9 Set up the parameters common to both the uplink and downlink.

Step 10 Set up the uplink specific parameters.

Step 11 Set up the downlink specific parameters. Note that when only a single transmit antenna is selected, that the transmit diversity scheme will be defaulted to the 'No Transmit Diversity (TD)' selection.

Step 12 Click the 'Continue' button. This causes the configuration window to close. A theoretical estimate for the selected system, using uncoded DS/QPSK, is displayed over the E_b/N_0 range in dB.

Step 13 (Optional) Click on the 'Clear' button to remove the estimated theoretical curve.

Step 14 Click on either the 'UPLINK Simulation' or 'DOWNLINK Simulation' button to start the simulation. The simulation continuous for each E_b/N_0 value specified in the 'E_b/N_0 range in dB' entry. Information on 'Simulation Completion' will be displayed. The plot will be updated as each simulation loop is completed. At completion of the simulation, a legend is added and results are displayed on the graph. At this stage, the result can be copied to the clipboard or retrieved from file.

Step 15 The main figure window can now be exited from by clicking on the 'Exit' button or alternatively, more simulations can be performed.

References

[1] K. Abend and B. D. Fritchman, "Statistical detection for communication channels with intersymbol interference," *Proceedings of the IEEE*, vol. 58, pp. 779–785, May 1970.

[2] M. Abramowitz and I. Stegun, *Handbook of Mathematical Functions*. Dover, 1972.

[3] E. Ackerman and A. Daryoush, "Broad-band external modulation fiber-optic links for antenna-remoting applications," *IEEE Trans. Microw. Theory Tech.*, vol. 45, pp. 1436–1442, August 1997.

[4] B. G. Agee, "The least-squares CMA: A new technique for rapid correction of constant modulus signals," in *Proc. IEEE ICASSP*, pp. 19.2.1–19.2.4, 1986.

[5] B. G. Agee, S. V. Schell, and W. A. Gardner, "Spectral self-coherence restoral: A new approach to blind adaptive signal extraction using antenna arrays," *Proc. IEEE*, vol. 78, pp. 753–767, November 1990.

[6] Z. Ahmed, D. Novak, R. Waterhouse, and H.-F. Liu, "Broad-band external modulation fiber-optic links for antenna-remoting applications," *IEEE Trans. Microw. Theory Tech.*, vol. 45, pp. 1431–1435, August 1997.

[7] P. D. Alexander, A. J. Grant and M. C. Reed, "Iterative detection in code-division multiple-access with error control coding," *European Trans. Telecommun.*, Vol. 9, July-August 1998.

[8] P. D. Alexander, L. K. Rasmussen, and C. B. Schlegel, "A linear receiver for coded multiuser CDMA," *IEEE Trans. Commun.*, vol. 45, pp. 605–610, May 1997.

[9] P. D. Alexander and L. K. Rasmussen, "On the windowed Cholesky factorisation of the time-varying asynchronous CDMA channel," *IEEE Trans. Commun.*, vol. 46, pp. 735–737, June 1998.

[10] F. Amaroso, "Use of DS/SS signaling to mitigate Rayleigh fading in a dense scatterer environment," *IEEE Pers. Commun. Mag.*, vol. 3, pp. 52–61, April 1996.

[11] S. Anderson, M. Millnert, M. Viberg, and B. Wahlberg, "An adaptive array for mobile communication systems," *IEEE Trans. Veh. Tech.*, vol. 40, pp. 230–236, February 1991.

[12] J. Anderson, T. Rappaport, and S. Yoshida, "Propagation measurements and models for wireless communications channels," *IEEE Commun. Mag.*, pp. 42–49, January 1995.

[13] D. Aszetly, *On Antennas Arrays in Mobile Communication Systems: Fast Fading and GSM Base Station Receiver Algorithm*. PhD thesis, Royal Institute of Technology, 1996.

[14] L. Bahl, J. Cocke, F. Jeinek, and J. Raviv, "Optimal decoding of linear codes for minimizing symbol error rate," in *Proc. IEEE ISIT*, p. 90, 1972.

[15] L. Bahl, J. Cocke, F. Jeinek, and J. Raviv, "Optimal decoding of linear codes for minimizing symbol error rate," *IEEE Trans. Inform. Theory*, vol. 20, pp. 248–287, March 1974.

[16] C. Balanis, *Antenna Theory - Analysis and Design*, USA: Wiley, seconded., 1997.

[17] C. Balanis, *Advanced Engineering Electromagnetics*. New York, NY: John Wiley & Sons, 1989.

[18] S. A. Barbulescu, *Iterative Decoding of Turbo Codes and other Concatenated Codes*. PhD thesis, University of South Australia, February 1996.

[19] M. Barrett and R. Arnott, "Adaptive antennas for mobile communications," *Electron. Commun. Eng. J.*, vol. 6, pp. 203–214, 1994.

[20] V. Barroso, M. Rendas, and J. Gomes, "Impact of array processing techniques on the design of mobile communication systems," in *Proc. IEEE MEC*, pp. 1291–1294, 1994.

[21] S. Benedetto and D. Divsalar, "Turbo codes: Comprehension, performance, analysis, design and iterative decoding." in *Proc. IEEE GLOBECOM Tutorial*, Sydney, Australia, November 1998.

[22] S. Benedetto and G. Montorsi, "Serial concatenation of block and convolutional codes," *Electr. Let.*, vol. 32, pp. 887–888, May 1996.

[23] S. Benedetto and G. Montorsi, "Performance of continuous and blockwise decoded turbo codes," *IEEE Commun. Let.*, vol. 1, pp. 77–79, May 1997.

[24] E. R. Berlekamp, *Algebraic Coding Theory*. McGraw-Hill: New York, 1968.

[25] E. Berlekamp, R. E. Pelle, and S. P. Pope, "The application of error control to communications," *IEEE Commun. Mag.*, vol. 25, pp. 44–57, April 1987.

[26] E. R. Berlekamp, R. E. Pelle, and S. P. Pope, "The application of error control to communications," *Proc. IEEE*, vol. 25, pp. 44–57, 1987.

[27] R. Bernhardt, "The use of multiple-beam directional antennas in wireless messaging systems," in *Proc. IEEE VTC*, pp. 858–861, 1995.

[28] F. Berens, A. Worm, H. Michel, and N. Wehn, "Implementation aspects of turbo-decoders for future radio applications," in *Proc. IEEE VTC*, (Amsterdam, The Netherlands), pp. 2601–2605, September 1999.

[29] C. Berrou, A. Clavieux, and P. Thitimasjshima, "Near Shannon limit error-correcting coding and decoding," in *Proc. IEEE ICC*, (Geneva, Switzerland), pp. 1064–1070, May 1993.

[30] H. Bertoni and G. Liang, "Review of ray modeling techniques for site specific propagation prediction," in *Wireless Communication - 'TDMA' vs. 'CDMA'* (S. Glisic and P. Leppänen, eds.), pp. 323–344, Kluwer Academic Publishers, 1997.

[31] T. E. Biedka, "Subspace constrained SCORE algorithms," in *Proc. Asilomar Conf. on Sign., Syst. Comp.*, pp. 716–720, 1993.

[32] A. Klein, J. J. Blanz and W. Mohr, "Measurement-based parameter adaptation of wideband spatial mobile radio channel models," in *Proc. IEEE ISSSTA*, pp. 91–97, 1996.

[33] F. Boutaud, *VLSI Signal Processing Solutions for Wireless Communications, a Technology Overview*, ch. 6, pp. 525–534. Kluwer, 1997.

[34] G. D. Boudreau, D. D. Falconer, and S. A. Mahmoud, "A comparison of trellis coded versus convolutionally coded spread-spectrum multiple-access systems," *IEEE Journ. Select. Areas Commun.*, vol. 8, pp. 628–640, May 1990.

[35] W. Braun and U. Dersch, "A physical mobile radio channel model," *IEEE Trans. Veh. Tech.*, vol. 40, pp. 472–482, May 1991.

[36] E. Buracchini, F. Muratore, V. Palestini, and M. Sinibaldi, "Performance analysis of a mobile system based on combined SDMA/CDMA access techniques," in *Proc. IEEE ISSSTA*, (Mainz, Germany), pp. 370–374, University of Kaiserslautern, 1996.

[37] D. Chase, "Digital signal design concepts for a time-varying Rician channel," *IEEE Trans. Commun.*, vol. 24, pp. 164–172, February 1976.

[38] K. L. Cheah, H. Sugimoto, T. J. Lim, L. K. Rasmussen, and S. Sumei, "Performance of hybrid interference canceller with zero-delay channel estimation for CDMA," in *Proc. IEEE GLOBECOM*, pp. 265–270, November 1998. Sydney, Australia.

[39] J. Cheung, M. Beach, and J. McGeehan, "Network planning for third-generation mobile radio systems," *IEEE Commun. Mag.*, vol. 32, pp. 54–59, November 1994.

[40] I. Chiba, W. Chujo, and M. Fujise, "Beam-space CMA adaptive array antennas," *Elect. Commun. Japan, Part 1*, vol. 78, no. 2, pp. 85–95, 1995.

[41] I. Chiba, W. Chujo, and Y. Karasawa, "Transmitting null beam forming with beam space adaptive array antennas," pp. 1498–1502, 1994.

[42] L. J. Cimini and N. R. Sollenberger, "OFDM with diversity and coding for high bit-rate mobile data applications," in *Proc. 3rd Intern. Workshop Mobile Multimedia Commun.*, p. A.3.1.1, September 1996.

[43] R. Clarke, "A statistical theory of mobile-radio reception," *Bell Sys. Tech. J.*, vol. 47, pp. 957–1000, 1968.

[44] G. C. Clark and J. Bibb Cain, *Error-Correction Coding for Digital Communications*. Plenum Press, 1988.

[45] I. Crohn and E. Bonek, "Modeling of intersymbol-interference in a Rayleigh fast fading channel with typical delay profile," *IEEE Trans. Veh. Tech.*, vol. 41, pp. 438–447, November 1992.

[46] T. M. Cover and J. A. Thomas, *Elements of Information Theory*, Wiley & Son, 1991.

[47] D. Cox, "Wireless network access for personal communications," *IEEE Commun. Mag.*, pp. 96–115, December 1992.

[48] D. Cox, R. Murray, and A. Norris, "800 MHz attenuation measured in and around suburban houses," *Bell Sys. Tech. J.*, vol. 673, pp. 921–954, July-August 1984.

[49] J. Craig, "A new, simple and exact result for calculating probability of two-dimensional signal constellations," in *Proc. IEEE MILCOM*, pp. 25.5.1–25.5.5, 1991.

[50] J. S. da Silva, B. Arroyo-Fernandez, B. Barani, J. Pereira, and D. Ikonomou, "Mobile and personal communications: ACTS and beyond," in *Wireless Communications: TDMA versus CDMA* (S. Glisic and P. Leppänen, eds.), pp. 379–414, Kluwer Academic Press, 1997.

[51] J. S. da Silva and B. Fernandes, "The European research program for advanced mobile systems," *IEEE Commun. Mag.*, vol. 33, pp. 14–19, February 1995.

REFERENCES   277

[52] D. Divsalar and F. Pollara, "Turbo codes for PCS applications," in *Proc. IEEE ICC*, (Seatle, WA), pp. 54–59, June 1995.

[53] P. Dent, B. Gudmundson and M. Ewerbring, "CDMA-IC: A novel code divisions multiple access scheme based on interference cancellation," in *Proc. IEEE PIMRC*, pp. 98–102, October 1992. Boston, Massachussets.

[54] C. Despins, D. Falconer, and S. Mahmoud, "Compound strategies of coding, equalization and space diversity for wideband TDMA indoor wireless channels," *IEEE Trans. Veh. Tech.*, vol. 41, pp. 369–379, 1992.

[55] D. Divsalar, F. Pollara, and R. J. McEliece, "Transfer function bounds on the performance of turbo codes," *JPL TDA Progress Report*, vol. 41, August 1995.

[56] D. Divsalar, M. Simon and D. Raphaeli, "Improved parallel interference cancellation for CDMA," *IEEE Trans. Commun.*, vol. 46, pp. 258–268, February 1998.

[57] A. Domingues, D. Caiado, N. Goncalves, and L. Correia, "Testing the COST231-WI propagation model in the city of Lisbon," Tech. Rep. COST 231 TD(96), European Cooperation in Field of Science and Technical Research, 1996.

[58] A. Duel-Hallen, "Decorrelating decision–feedback multiuser detector for synchronous code–division multiple–access channel," *IEEE Trans. Commun.*, vol. 41, pp. 285–290, February 1993.

[59] G. Efthymoglou, V. Aalo, and H. Helmken, "Performance analysis of coherent DS-CDMA systems in a Nakagami fading channel with arbitrary parameters," *IEEE Trans. Veh. Tech.*, vol. 46, pp. 289–297, May 1997.

[60] P. Eggers, "TSUNAMI: Spatial radio spreading as seen by directive antennas," Tech. Rep. COST 231 TD(94) 119, EURO-COST, September 1994.

[61] H. Elders-Boll, H. D. Schotten, and A. Busboom, "Efficient implementation of linear multiuser detectors for asynchronous CDMA systems by linear interference cancellation," *European Trans. Telecommun.*, Vol. 9, No. 4, pp. 427–437, September-November 1998.

[62] T. Eng and L. Milstein, "Coherent DS-CDMA performance in Nakagami multipath fading," *IEEE Trans. Commun.*, vol. 43, pp. 1134–1143, February/March/April 1995.

[63] R. B. Ertel, "Vector Channel Model Evaluation." Tech. Report, SW Bell Tech. Resources, August 1997.

[64] R. Ertel, K. Sowerby, T. Rappaport, and J. Reed, "Overview of spatial channel models for antenna array communication systems," *IEEE Pers. Commun. Mag.*, vol. 5, pp. 10–22, February 1998.

[65] The ETSI UMTS Terrestrial Radio Access (UTRA) ITU-R RTT Candidate Submission, *ETSI document SMG2 260/98*.

[66] UMTS 30.01 Baseline Document. *ETSI SMG #23*, Budapest, Hungary, October 1997.

[67] UMTS 30.03 version 3.2.0, Annex B, ETSI TR 101 112 V3.2.0 (1998-04).

[68] UTRA Physical Layer Description - FDD Parts, v0.4, 1998-06-25. ETSI document SMG2 UMTS-L1 221/98.

[69] E. A. Fain and M. K. Varanasi, "Signal design to optimize the total capacity of multiuser decoders under location-invariant bandwidth constraints," in *Proc. IEEE GLOBECOM: Communications Theory Mini Conference*, (Sydney, Australia), pp. 100–105, November 1998.

[70] C. Farsakh and J. Nossek, "Comparison of symmetric antenna array configurations by their spatial separation potential," in *Proc. IEEE ICUPC*, pp. 433–436, 1996.

[71] K. Feher, "MODEMS for emerging digital cellular-mobile radio systems," *IEEE Trans. Veh. Tech.*, vol. 40, pp. 355–365, May 1991.

[72] L. Felsen and N. Marcuvitz, *Radiation and Scattering of Waves*. Englewood Cliffs, NJ: Prentice-Hall, Inc., 1973.

[73] J. R. Fonollosa, J. A. R. Fonollosa, Z. Zonar, and J. Vidal, "Blind multiuser identification and detection in CDMA systems," in *Proc. IEEE ICASSP*, (Detroit, Michigan), pp. 1876-1879, 1995.

[74] G. D. Forney, "The Viterbi Algorithm," *Proc. IEEE*, pp. 268–278, March 1973.

[75] G. Foschini, "Layered space-time architecture for wireless communication in a fading environment when using multi-element antennas," *Bell Sys. Tech. J.*, vol. 1, 1996.

[76] G. J. Foschini and M. J. Gans, "Capacity when using diversity at transmit and receive sites and the Rayleigh-faded matrix channel is unknown at the transmitter," in *WINLAB Workshop on Wireless Information Networks*, (New Brunswick, New Jersey), March 1996.

[77] S. Fortune, D. Gay, B. Kernighan, and et. al., "WISE Design of Indoor Wireless Systems: practical computation and optimization," *IEEE Comp. Sc. Eng.*, pp. 58–68, 1995.

[78] T. Freeburg, "Enabling technologies for wireless in-building network communications - four technical challenges, four solutions," *IEEE Commun. Mag.*, pp. 58–64, April 1991.

[79] P. K. Frenger, P. Orten and T. Ottosson, "Code-spread CDMA with interference cancellation," to appear in *IEEE Journ. Select. Areas Commun.*

[80] P. Frenger, P. Orten, and T. Ottosson, "Combined coding and spreading in CDMA systems using maximum free distance convolutional codes," in *Proc. IEEE VTC*, (Ottawa, Canada), May 1998.

[81] P. Frenger, P. Orten, and T. Ottosson, "Code-spread CDMA using low-rate convolutional codes," in *Proc. IEEE ISSSTA*, (Sun City, South Africa), pp. 374–378, September 1998.

[82] P. K. Frenger, P. Orten, T. Ottosson and A. B. Svensson, "Rate-compatible convolutional codes for multirate DS-CDMA systems," *IEEE Trans. Commun.*, ,vol. 47, pp. 828–836, June 1999.

[83] J. Fuhl, A. Molisch and E. Bonek, "Unified channel model for mobile radio systems with smart antennas," *IEE Proc.-Radar, Sonar Navig.*, vol. 145, pp. 32–41, Febraury 1998.

[84] M. Fujimoto, N. Kikuma, and N. Inagaki, "Performance of CMA adaptive array optimized by the Marquardt method for suppressing multipath waves," *Elect. Commun. Japan, Part 1*, vol. 75, no. 9, pp. 89–100, 1992.

[85] M. Gagnaire, "An overview of broad-band access technologies," *Proc. IEEE*, vol. 85, pp. 1958–1972, December 1997.

[86] R. G. Gallager, *Information Theory and Reliable Communication*. Wiley: New York, 1968.

[87] T. R. Giallorenzi and S. G. Wilson, "Multiuser ML sequence estimator for convolutionally coded asynchronous DS-CDMA systems," *IEEE Trans. Commun.*, vol. 44, pp. 997–1008, August 1996.

[88] K. Gilhousen, I. Jacobs, R. Padovani, A. Viterbi, L. W. Weaver Jr. and C. E. Wheatly III, "On the capacity of a cellular CDMA system," *IEEE Trans. Veh. Tech.*, vol. 40, pp. 303–311, May 1991.

[89] L. Godara, "Applications of antenna arrays to mobile communications, Part I: Performance Improvement, Feasibility and System Considerations," in *Proc. IEEE*, vol. 85, pp. 1031–1060, 7 1997.

[90] M. Goldburg and R. Roy, "The impacts of SDMA on PCS system design," (San Diego, CA), pp. 242–246, 1994.

[91] G. H. Golub and C. F. V. Loan, *Matrix Computations*. The John Hopkins University Press, 2nd ed., 1989.

[92] R. Gooch and J. Lundell, "The CM array: An adaptive beamformer for constant modulus signals," in *Proc. IEEE ICASSP*, pp. 1523–2526, 1986.

[93] A. J. Grant and P. D. Alexander, "Random sequence multisets for synchronous code-division multiple-access channels," *IEEE Trans. Inform. Theory*, vol. 44, pp. 2832–2836, November 1998.

[94] P. M. Grant, "Algorithms and Antenna Array Recommendations." Tech. Report, AO20/AUC/A12DR/P1/xx-D2.1.2, Tsunami (II), September 1996.

[95] P. Grant, J. Thompson, and B. Mulgrew, "Antenna arrays for cellular CDMA systems," in *CDMA techniques for 3rd generation mobile systems* (F. Swarts, P. van Rooyen, I. Oppermann, and M. Lötter, eds.), ch. 5, Kluwer Academic Publihers, 1998.

[96] L. J. Greenstein, "Microcells in personal communicaitons systems," *IEEE Commun. Mag.*, pp. 76–88, December 1992.

[97] L. Greenstein, V. Erceg, Y. S. Yeh, and M. Clark, "A new path-gain/delay-spread propagation model for digital cellular channels," *IEEE Trans. Veh. Tech.*, vol. 46, pp. 477–485, May 1997.

[98] D. Guo, L. K. Rasmussen, and T. J. Lim, "Linear parallel interference cancellation in random-code CDMA," to appear in *IEEE Journ. Select. Areas Commun.*

[99] L. A. Hageman and D. M. Young, *Applied iterative methods*. Academic Press, 1981.

[100] J. Hagenauer, "Iterative decoding of binary block and convolutional codes," *IEEE Trans. Inform. Theory*, vol. 42, pp. 429–445, March 1996.

[101] E. Hall and S. G. Wilson, "Design and analysis of turbo codes on Rayleigh fading channels," *IEEE Journ. Select. Areas Commun.*, vol. 16, pp. 160–174, February 1998.

[102] T. Hattori and K. Hirade, "Multi-transmitter simulcast digital signal transmission by using frequency offset strategy in land mobile radiotelephone system," *IEEE Trans. Veh. Tech.*, vol. VT-27, pp. 170–176, May 1992.

[103] S. Haykin, *Digital Communications*. Wiley, 1988.

[104] S. Haykin, *Adaptive Filter Theory*. Englewood Cliffs, New Jersey: Prentice Hall, 1991.

[105] T. Hentschel and G. Fettweis, *Software radio receivers*, in *CDMA techniques for 3rd generation mobile systems* (F. Swarts, P. van Rooyen, I. Oppermann, and M. Lötter, eds.), ch. 10, Kluwer Academic Publihers, 1998. Kluwer, 1998.

[106] K. Hilal and P. Duhamel, "A blind equalizer allowing soft transition between the constant modulus and decision-directed algorithm for PSK modulated signals," in *Proc. IEEE ICC*, pp. 1144–1148, 1993.

[107] A. Hiroike, F. Adachi, and N. Nakajima, "Combined effects of phase sweeping transmitter diversity and channel coding," *IEEE Trans. Veh. Tech.*, vol. 41, pp. 170–176, May 1992.

[108] M. Ho, G. L. Stüber, and M. D. Austin, "Performance of switched-beam smart antennas for cellular radio systems," *IEEE Trans. Veh. Tech.*, vol. 47, pp. 10–19, February 1998.

[109] W. Honcharenko, H. Bertoni, J. Dailing, and et. al., "Mechanisms governing UHF propagation on single floors in modern office buildings," *IEEE Trans. Veh. Tech.*, vol. 41, pp. 496–504, November 1992.

[110] D. Hong and S. Rappaport, "Traffic model and performance analysis for cellular mobile radio telephone systems with prioritized and non-prioritized handoff procedures," *IEEE Trans. Veh. Tech.*, vol. VT-35, pp. 77–92, August 1986.

[111] A. Hottinen and R. Wichman, "Transmit diversity by antenna selection in CDMA downlink," in *Proc. IEEE ISSSTA*, (Sun City, South Africa), pp. 767–770, September 1998.

[112] J. Hui, "Throughput analysis for code division multiple accessing of the spread spectrum channel," *IEEE Journ. Select. Areas Commun.*, vol. SAC-2, July 1984.

[113] F. Ikegami and S. Yoshida, "Analysis of multipath propagation structure in urban mobile radio environments," *IEEE Trans. Ant. Prop.*, vol. AP-28, pp. 531–537, July 1980.

[114] F. Ikegami, S. Yoshida, T. Takeuchi, and M. Umehira, "Propagation factors controlling mean field strength on urban streets," *IEEE Trans. Ant. Prop.*, vol. AP-32, pp. 822–829, August 1984.

[115] W. Jakes, *Microwave Mobile Communications*. Wiley: New York, 1974.

[116] K. Jamal and E. Dahlman, "Multi-stage serial interference cancellation for DS-CDMA," in *Proc. IEEE VTC '96*, pp. 671–675, April 1996. Atlanta, Georgia.

[117] N. Jayant, "Signal compression: Technology targets and research directions," *IEEE Journ. Select. Areas in Commun.*, vol. 10, pp. 796–818, June 1992.

[118] I. Jeong and M. Nakagawa, "A novel transmission diversity system in TDD-CDMA," in *Proc. IEEE ISSSTA*, (Sun City, South Africa), pp. 771–775, September 1998.

[119] A.-L. Johansson and A. Svensson, "On multi-rate DS/CDMA schemes with interference cancellation," to appear in *Journ. Wireless Personal Commun.*, Kluwer, Vol. 9, No. 1, pp. 1-29, January 1999.

[120] M. Jones and M. A. Wickert, "Direct sequence spread spectrum using directionally constrained adaptive beamforming to null interference," *IEEE Journ. Select. Areas Commun.*, vol. 13, pp. 71-79, January 1995.

[121] P. Jung, P. Baier, and A. Steil, "Advantages of CDMA and spread spectrum techniques over FDMA and TDMA in cellular mobile radio applications," *IEEE Trans. Veh. Tech.*, vol. 42, pp. 357-364, August 1993.

[122] P. Jung, B. Steiner, and B. Stilling, "Exploitation of intracell macrodiversity in mobile radio systems by deployment of remote antennas," in *Proc. IEEE ISSSTA*, (Mainz,Germany), pp. 302-307, University of Kaiserslautern, 1996.

[123] M. J. Juntti and B. Aazhang, "Finite memory-length linear multiuser detection for asynchronous CDMA communications," *IEEE Trans. Commun.*, vol. 45, pp. 611-622, May 1997.

[124] M. J. Juntti, B. Aazhang, and J. O. Lilleberg, "Iterative implementation of linear multiuser detection for dynamic asynchronous CDMA systems," *IEEE Trans. Commun.*, vol. 46, pp. 503-508, April 1998.

[125] M. Kawabe, S. Sato, H. Sugimoto, T. Sato and A. Fukasawa, "Interference Cancellation system using estimations of propagation parameters," in *Proc. JTC-CSCC*, pp. 173-178, 1994. Japan.

[126] J. Keller, "Geometrical theory of diffraction," *J. Optical Soc. America*, vol. 52, pp. 116-130, February 1962.

[127] C. Kermarrec, *RFIC design for wireless communication*, ch. 5, pp. 509-523. Kluwer, 1997.

[128] B. H. Khalaj, A. Paulraj, and T. Kailath, "2D RAKE receivers for CDMA cellular systems," in *Proc. IEEE GLOBECOM*, pp. 400-404, 1994.

[129] A. Klein and W. Mohr, "A statistical wideband mobile radio channel model including the direction of arrival," in *Proc. IEEE ISSSTA*, pp. 102-06, 1996.

[130] R. Kohno, H. Imai and M. Hatori, " Cancellation techniques for co-channel interference in asynchronous spread spectrum multiple access systems", *Elect. Commun.*, Vol. 66, pp 20-29, May 1983.

[131] R. Kohno, H. Imai, M. Hatori, and S. Pasupathy, "Combination of an adaptive array antenna and a canceller of interference for direct-sequence spread-spectrum multiple-access system," *IEEE Journ. Select. Areas Commun.*, vol. 8, pp. 675-681, May 1990.

[132] R. Kohno, "Spatially and temporally joint optimum transmitter - receiver based on adaptive array antenna for multi-user detection in DS/CDMA," in *Proc. IEEE ISSSTA*, (Mainz, Germany), pp. 365–369, University of Kaiserslautern, 1996.

[133] M. R. Koohrangpour and A. B. Svensson, "Joint interference cancellation and Viterbi decoding in DS-CDMA," in *Proc. IEEE PIMRC*, pp. 1161–1165, September 1997, Helsinki, Finland.

[134] R. Kouyoumjian and P. Pathak, "A uniform geometrical theory of diffraction for an edge in a perfectly conducting surface," *Proc. IEEE*, vol. 62, pp. 1448–1461, November 1974.

[135] H. Krim and M. Viberg, "Two decades of array signal processing research - the parametric approach," *IEEE Signal Proc. Mag.*, vol. 13, pp. 67–94, July 1996.

[136] F. R. Kschischang and B. J. Frey, "Iterative decoding of compound codes by probability propagation in graphical models," *IEEE Journ. Select. Areas Commun.*, vol. 2, pp. 219–230, February 1998.

[137] W. Y. Kuo and M. P. Fitz, "Design and analysis of transmitter diversity using intentional frequency offset for wireless communications," *IEEE Trans. Veh. Tech.*, vol. VT-46, pp. 691–698, November 1997.

[138] P. J. Lanzkron, D. J. Rose, and D. B. Szyld, "Convergence of nested classical iterative methods for linear systems," *Numer. Math.*, vol. 58, pp. 685–702, 1991.

[139] W. C. Y. Lee, "Antenna spacing requirements for a mobile base-station diversity," *Bell Sys. Tech. J.*, vol. 50, pp. 1859–1877, July-August 1971.

[140] W. C. Y. Lee, *Mobile Communication Engineering*. McGraw Hill Publications, 1982.

[141] W. Lee, "Overview of cellular CDMA," *IEEE Trans. Veh. Tech.*, vol. 40, pp. 291–302, May 1991.

[142] W. Lee, *Mobile Communications Design Fundamentals*. Wiley Series in Telecommunications, 1993.

[143] W. C. Y. Lee, *Mobile Cellular Telecommunications*. McGraw Hill Publications, 1995.

[144] J. S. Lee and L. E. Miller, *CDMA Systems Engineering Handbook*. Artech House Publishers, 1998.

[145] K. Leung, W. Massey, and W. Whitt, "Traffic models for wireless communication networks," *IEEE Journ. Select. Areas Commun.*, vol. 12, pp. 1353–1364, October 1994.

[146] G. Liang and H. Bertoni, "Review of ray modelling techniques for site specific propagation prediction," in *Wireless Communications: TDMA versus CDMA* (S. Glisic and P. Leppänen, eds.), Kluwer Academic Publishers, 1997.

[147] J. C. Liberti, *Analysis of CDMA Cellular Radio Systems Employing Adaptive Antennas.* PhD thesis, Virginia Polytechnic Inst. and State Uni., 1995.

[148] J. C. Liberti and T. S. Rappaport, "A geometrically based model for line of sight multi-path radio channels," in *Proc. IEEE VTC*, pp. 844–48, 1996.

[149] J. C. Liberti and T. S. Rappaport, *Smart Antennas for Wireless Communications: IS-95 and Third Generation CDMA Applications*, Prentice Hall, 1999.

[150] T. J. Lim, D. Guo, and L. K. Rasmussen, "Noise enhancement in the family of decorrelating detectors for multiuser CDMA," in *Proc. IEEE S'pore Int. Conf. on Commun. Systems*, November 1998. Singapore.

[151] T. J. Lim, L. K. Rasmussen, and H. Sugimoto, "The effects of phase offset on the performance of binary and quaternary phase shift keying in multiuser CDMA system," in *Proc. IEEE ICT*, vol. II, pp. 42–96, IEEE, 1998.

[152] T. J. Lim, L. K. Rasmussen, and H. Sugimoto, "Performance of CDMA systems in the presence of complex valued multiuser interference," *Kluwer Academic Publishers*, To be published 2000.

[153] T. J. Lim and S. Roy, "Adaptive filters in multiuser (MU) CDMA detection," *Wireless Networks.*, vol. 4, no. 4, pp. 307–318, 1998.

[154] S. Lin and D. J. Costello, Jr, *Error control coding - Fundamentals and Applications.* Prentice-Hall, 1983.

[155] A. R. Lopez, "Performance predictions for cellular switched-beam intelligent antenna systems," *IEEE Commun. Mag.*, pp. 152–154, October 1996.

[156] M. Lötter, F. Swarts, and P. van Rooyen, "CDMA technology: Changing the face of wireless access," *Trans. South African Inst. Elect. Eng.*, vol. 89, pp. 89–91, September 1998.

[157] M. Lötter, *Numerical analysis of Spatial/Temporal cellular CDMA.* PhD thesis, University of Pretoria, 1999.

[158] M. Lötter and P. van Rooyen, "Performance of DS/CDMA systems with antenna arrays in non-uniform environments," *IEEE Journ. Select. Areas Commun.*, 1999. Accepted for publication.

REFERENCES 285

[159] M. Lötter and P. van Rooyen, "Modeling spatial aspects of cellular CDMA/SDMA systems.," *IEEE Commun. Let.*, vol. 3, pp. 128–131, May 1999.

[160] K. Löw, "Comparison of urban propagation models with CW measurements," in *Proc. IEEE VTC*, vol. 2, pp. 96–942, IEEE, 1992.

[161] J. Lu, K. B. Letaief, M. L. Liou, and J. C.-I. Chuang, "On the use of modulation and diversity for enhancing cellular spectrum efficiency for wireless multimedia communication networks," in *Proc. IEEE GLOBECOM*, (Phoenix, Arizona, USA), pp. 1168–1172, November 1997.

[162] M. Lu, T. Lo, and J. Litva, "A physical spatio-temporal model of multipath propagation channels," in *Proc. IEEE VTC*, pp. 180–184, 1997.

[163] R. Lucky, "New communications services - what does society want?," *Proc. IEEE*, vol. 85, pp. 1536–1543, October 1997.

[164] R. W. Lucky, J. Salz, and E. J. Weldon, *Principles of Data Communication*. New York, NY: McGraw-Hill, 1968.

[165] R. Luebbers, "Finite conductivity uniform GTD versus knife edge diffraction in prediction of propagation path loss," *IEEE Trans. Ant. Prop.*, vol. AP-32, pp. 70–76, January 1984.

[166] H. Lui and M. Zoltowski, "Blind equalization in antenna array CDMA systems," *IEEE Trans. Signal Proc.*, vol. 45, pp. 161–172, January 1997.

[167] R. Lupas and S. Verdú, "Near-far resistance of multiuser detectors in asynchronous channels," *IEEE Trans. Commun.*, vol. 38, pp. 496–508, April 1990.

[168] M. Lyons, F. Burton, B. Egan, T. Lynch, and S. Skelton, "Dynamic modeling of present and future service demand," *Proc. IEEE*, vol. 85, pp. 1544–1555, October 1997.

[169] U. Madhow and M. Honig, "MMSE Interference Supression for Direct Sequence Spread Spectrum CDMA " *IEEE Trans. Commun.*, Vol. 42, No. 12, December 1994. pp 3178-3188.

[170] T. L. Marzetta and B. M. Hochwald, "Capacity of a mobile multipleantenna communication link in Rayleigh flat fading," *IEEE Trans. Inform. Theory*, pp. 139–157, vol. 45, January 1999.

[171] A. Mathur, A. V. Keerthi, and J. J. Shynk, "Estimation of correlated cochannel signals using the constant modulus array," in *Proc. IEEE ICC*, pp. 1525–1529, 1995.

[172] A. Mathur, A. V. Keerthi, and J. J. Shynk, "Co-channel signal recovery using the MUSIC algorithm and the constant modulus array," *IEEE Sign. Proc. Let.*, vol. 2, pp. 191–194, October 1995.

[173] S. Miller, "An adaptive direct-sequence code-division multiple-access receiver for multi-user interference rejection", *IEEE Trans. Commun.*, Vol. 43, No. 2/3/4, February/March/April 1995. pp 1746-1755.

[174] M. Mizuno and T. Ohgane, "Application of adaptive array antennas to radio communications," *Electron. Commun. Japan*, vol. 77, pp. 48–59, 1994.

[175] W. Mohr, "Impact of radio channel properties on adaptive antenna concepts for terrestrial mobile radio applications," in *Proc. IEEE ICT*, (Melbourne, Sydney), pp. 405–410, 2-5 April 1997.

[176] J. Moss, D. Edwards, and K. Allen, "Radio imaging for the validation of propagation predictions," in *Proc. IEEE PIMRC*, pp. 505–508, September 1-4 1997.

[177] P. Mundra, T. Singai, and R. Kapur, "The choice of a digital modulation scheme in a mobile radio system," in *Proc. IEEE VTC*, pp. 1–4, 1993.

[178] M. Nakagami, "The m-distribution - a general formula of intensity distribution of rapid fading," in *Statistical Methods in Radio Wave Propagation* (W. Hoffman, ed.), Elmsford, NY: Pergamon, 1960.

[179] A. Naguib, A. Paulraj, and T. Kailath, "Capacity improvement with base-station antenna arrays in cellular CDMA," *IEEE Trans. Veh. Tech.*, vol. 43, pp. 691–698, August 1994.

[180] A. F. Naguib, V. Tarokh, N. Seshadri, and A. R. Calderbank, "A space-time coding modem for high-data-rate wireless communications," *IEEE Journ. Select. Areas Commun.*, vol. 16, pp. 1459–1478, October 1998.

[181] S. Nordholm, I. Claesson, and B. Bengtsson, "Adaptive array noise suppression of handsfree speaker input in cars," *IEEE Trans. Veh. Tech.*, vol. 42, pp. 514–518, November 1993.

[182] O. Norklit and J. B. Anderson, "Mobile radio environments and adaptive arrays," in *Proc. IEEE PIMRC*, pp. 725–28, 1994.

[183] O. Norklit, P. Eggers, P. Zetterberg, and J. B. Andersen, "The angular aspects of wideband modeling and measurements," in *Proc. IEEE ISSSTA*, (Mainz, Germany), pp. 73–78, University of Kaiserslautern, 1996.

[184] T. Ohgane, T. Shimura, N. Matsuzawa, and H. Sasaoka, "An implementation of CMA adaptive array for high speed GMSK transmission in mobile communications," *IEEE Trans. Veh. Tech.*, vol. 42, pp. 282–288, August 1993.

[185] T. Ojanperä, "Overview of research activities for third generation mobile communication," in *Wireless Communications - TDMA versus CDMA*

(S. Glisic and P. Leppänen, eds.), pp. 415–446, Kluwer Academic Press, 1997.

[186] K. Okada and F. Kubota, "On the dynamic channel assignment in small zone systems for mobile communications," *Rev. CRL*, vol. 36, pp. 113–123, June 1990.

[187] Y. Okumura, E. Ohmori, T. Kwano, and K. Fukuda, "Field strength and its variability in VHF and UHF land-mobile radio services," *Review Elect. Commun. Lab.*, pp. 826–873, 1968.

[188] H. Olofsson, M. Almgren, and M. Hook, "Transmitter diversity with antenna hopping for wireless communication networks," in *Proc. IEEE VTC*, (Phoenix, Arizona, USA), May 1997.

[189] B. Ottersten, "Spatial Division Multiple Access (SDMA) in Wireless Communications," in *Proc. Nordic Radio Symposium*, 1995.

[190] T. Oyama, S. Sun, H. Sugimoto, T. J. Lim, L. K. Rasmussen, and Y. Matsumoto, "Performance comparison of multi-stage SIC and limited treesearch detection in CDMA," in *Proc. IEEE VTC*, pp. 1854–1858, May 1998. Ottawa, Canada.

[191] I. Parra, G. Xu, and H. Lui, "A least squares projective constant modulus approach," in *Proc. IEEE PIMRC*, pp. 673–676, 1995.

[192] J. Parsons, *The Mobile Radio Propagation Channel*. McGraw Hill: New York, 1992.

[193] P. Patel and J. Holtzman, "Analysis of simple successive interference cancellation scheme in a DS/CDMA," *IEEE Journ. Select. Areas Commun.*, vol. 12, pp. 796–807, June 1994.

[194] A. Paulraj and C. Papadias, "Space-time processing for wireless communications," *IEEE Sign. Proc. Mag.*, pp. 49–83, November 1997.

[195] A. Paulraj, R. Roy, and T. Kailath, "Estimation of signal parameters by rotational invariance techniques (ESPRIT)," in *Proc. Asilomar Conf. on Sign., Syst. Comp.*, 1985.

[196] K. Pehkonen and P. Komulainen, "A superorthogonal turbo-code for CDMA applications," in *ISSSTA '96: International Symposiun on Spread Spectrum Techniques & Apllications*, (Mainz, Germany), pp. 580–584, September 1996.

[197] K. Pehkonen and P. Komulainen, "Performance evaluation of superorthogonal turbo codes in AWGN and flat Rayleigh fading channels," *IEEE Journ. Select. Areas Commun.*, vol. 16, pp. 196–205, February 1998.

[198] W. W. Peterson and E. J. Weldon, *Error-Correcting Codes*. Cambridge Mass.: MIT Press, 1972.

[199] J. H. Reed, P. Petrus and T. S. Rappaport, "Geometrically based statistical channel model for macrocellular mobile environments," in *Proc. IEEE GLOBECOM*, pp. 1197–1201, 1996.

[200] P. Petrus, *Novel Adaptive Array Algorithm and their Impact on Cellular System Capacity*. PhD thesis, Virginia Polytechnic Inst. and State Uni., 1997.

[201] R. Pickholtz and K. Elbarbary, "The recursive constant modulus algorithm: A new approach for real-time array processing," in *Proc. Asilomar Conf. on Sign., Syst. Comp.*, pp. 627–632, 1993.

[202] G. J. Pottie, "System design issues in personal communications," *IEEE Pers. Commun. Mag.*, vol. 2, pp. 50–67, October 1995.

[203] R. Prasad, "An overview of multi-carrier CDMA," in *Proc. IEEE ISSSTA*, (Mainz, Germany), pp. 107–114, University of Kaiserlautern, September 1996.

[204] R. Prasad, *CDMA for Wireless Personal Communications*. Artech House, 1996.

[205] J. G. Proakis, *Digital Communications*. McGraw–Hill, 3rd ed., 1995.

[206] M. Pursley, "Performance evaluation for phase-coded spread-spectrum multiple access communications - Part I: system analysis," *IEEE Trans. Commun.*, vol. COM-25, pp. 795–799, August 1977.

[207] K. Raith and J. Uddenfeldt, "Capacity of digital cellular TDMA systems," *IEEE Trans. Veh. Tech.*, vol. 40, pp. 323–331, May 1991.

[208] P. Rapajic and B.S. Vucetic, "Adaptive receiver structures for asynchronous CDMA systems", *IEEE Journ. Select. Areas Commun.*, Vol. 12, No. 4, May 1994. pp 685-697.

[209] T. Rappaport, *Wireless Communications, Chps. 3 & 4*. Upper Saddle River, NJ: Prentice Hall, 1996.

[210] F. Rashid-Farrokhi, K. R. Liu, and L. Tassiulas, "Transmit beamforming and power control for cellular wireless systems," *IEEE Journ. Select. Areas Commun.*, vol. 16, pp. 1437–1450, October 1998.

[211] F. Rashid-Farrokhi, L. Tassiulas, and K. J. R. Liu, "Joint optimal power control and beamforming for wireless networks with antenna arrays," in *Proc. IEEE GLOBECOM*, 1996.

[212] F. Rashid-Farrokhi, L. Tassiulas, and K. J. R. Liu, "Joint optimal power control and beamforming in wireless networks with antenna arrays," *IEEE Trans. Comm.*, 1998.

[213] G. Raleigh and J. M. Cioffi, "Spatio-temporal coding for wireless communications," in *Proc. IEEE GLOBECOM*, pp. 1809–1814, December 1996.

[214] G. G. Raleigh and A. Paulraj, "Time varying vector channel estimation for adaptive spatial equalization," in *Proc. IEEE GLOBECOM*, pp. 218–24, 1995.

[215] L. K. Rasmussen, T. J. Lim, and T. M. Aulin, "Breadth-first maximum-likelihood detection in multiuser CDMA," *IEEE Trans. Commun.*, vol. 45, pp. 1176–1178, October 1997.

[216] L. K. Rasmussen, P. D. Alexander, and T. J. Lim, "A linear model for CDMA signals received with multiple antennas over multi-path fading channels," in *CDMA techniques for 3rd generation mobile systems* (F. Swarts, P. van Rooyen, I. Oppermann, and M. Lötter, eds.), ch. 2, Kluwer Academic Publihers, 1998.

[217] J. Razavilar, F. Rashid-Farrokhi, and K. J. R. Liu, "Distributed co-channel interference control in cellular radio systems," *IEEE Trans. Veh. Tech.*, 1992.

[218] J. Razavilar, F. Rashid-Farrokhi, and K. J. R. Liu, "Software radio architecture with smart antennas: a tutorial on algorithms and complexity," *IEEE Journ. Select. Areas Commun.*, 1999.

[219] M. Reed, *Iterative Receiver Techniques for Coded Multiple Access Communication Systems.* Ph.D. Dissertation, University of South Australia, November 1998.

[220] M. C. Reed, C. B. Schlegel, P. D. Alexander and J. A. Asenstorfer, "Iterative multiuser detection for CDMA with FEC: Near-single-user performance," *IEEE Trans. Commun.*, vol. 46, pp. 1693–1699, December 1998.

[221] "A Regulatory Framework for UMTS." Report #1 from the UMTS Forum, June 1997.

[222] K. Rikkinen, "Comparison of very low rate coding methods for CDMA radio communications systems," in *Proc. IEEE ISSSTA*, (Oulu, Finland), pp. 268–272, July 1994.

[223] M. Y. Rhee, *Error-Correcting Coding Theory.* McGraw-Hill, 1989.

[224] P. Robertson, "Improving decoder and code structure of parallel concatenated recursive systematic (turbo) codes," in *Proc. IEEE ICUPC*, pp. 183–187, 1994.

[225] P. Robertson, "Illuminating the structure of parallel concatenated recursive systematic (turbo) codes," in *Proc. IEEE GLOBECOM*, pp. 1298–1303, 1994.

[226] Z. Rong, "Simulation of adaptive array algorithms for CDMA systems," Masters Thesis, Virginia Polytechnic Institute and State University, 1996.

[227] Z. Rong, P. Petrus, T. S. Rappaport, and J. H. Reed, "Despread-respread multi-target constant modulus array for CDMA Systems," *IEEE Commun. Let.*, vol. 1, pp. 114–116, July 1997.

[228] Z. Rong, T. S. Rappaport, P. Petrus, and J. H. Reed, "Simulation of multi-target adaptive array algorithms for wireless CDMA systems," in *Proc. IEEE VTC*, pp. 1–5, 1997.

[229] J. Rossi and A. Levi, "A ray model for decimetric radiowave propagation in an urban area," *Radio Science*, vol. 27, no. 6, pp. 971–79, 1993.

[230] M. J. Rude and L. J. Griffiths, "Incorporation of linear constraints into the constant modulus algorithm," in *Proc. IEEE ICASSP*, pp. 968–971, 1989.

[231] A. M. Saleh and R. A. Valenzuela, "A statistical model for indoor multipath propagation," *IEEE Journ. Select. Areas Commun.*, vol. 5, Feburary 1987.

[232] J. Salz and J. H. Winters, "Effects of correlated fading on adaptive arrays in digital mobile radio," *IEEE Trans. Veh. Tech.*, vol. 43, pp. 1049–1057, November 1994.

[233] C. Sankaran and A. Ephremides, "Sovling a class of optimum multiuser detection problems with polynomial complexity," *IEEE Trans. Inform. Theory*, vol. 44, pp. 1958–1961, September 1998.

[234] M. Sawahashi, H. Andoh, and K. Higuchi, "Interference rejection weight control for pilot symbol-assisted coherent multistage interference canceller using recursive channel estimation in DS-CDMA mobile radio," *IEICE Trans. Fundamentals*, vol. E81-A, pp. 957–970, May 1998.

[235] M. Sawahashi, Y. Miki, H. Andoh and K. Higuchi, "Serial canceler using channel estimation by pilot symbols for DS-CDMA," *IEICE Tech. Rep. No. RCS95-50*, vol. 12, pp. 43–48, July 1995.

[236] L. L. Scharf, *Statistical Signal Processing*. Addison Wesley, 1990.

[237] K. R. Schaubach, N. J. Davis, and T. S. Rappaport, "A ray tracing method for predicting path loss and delay spread in microcellular environment," in *Proc. IEEE VTC*, pp. 932–35, 1992.

[238] C. Schlegel, *Trellis Coding*, IEEE Press, 1997.

[239] R. O. Schmidt, *A Signal Subspace Approach to Multiple-emitter Location and Spectral Estimation*. PhD thesis, Stanford University, 1981.

[240] R. O. Schmidt, "Multiple emitter location and signal parameter estimation," *IEEE Trans. Ant. Prop.*, vol. AP-34, pp. 276–280, March 1986.

[241] S. Y. Seidel and T. S. Rappaport, "Site-specific propagation prediction for wireless in-building personal communication system design," *IEEE Trans. Veh. Tech.*, vol. 43, November 1994.

[242] N. Seshadri and J. H. Winters, "Two signalling schemes for improving the error performance of frequency-division-duplex transmission system using transmitter antenna diversity," *Intern. J. Wireless Inf. Netw.*, vol. 1, pp. 49–60, 1994.

[243] N. Seshadri, V. Tarokh, and A. R. Calderbank, "Space-time codes for wireless communications: Code construction," in *Proc. IEEE VTC*, pp. 637–641, 1997.

[244] Q. S. Spencer, "A statistical model for angle of arrival in indoor multipath propagation," in *Proc. IEEE VTC*, pp. 1415–19, 1997.

[245] M. Shafi, A. Hashimoto, M. Umehira, S. Ogose, and T. Murase, "Wireless communications in the twenty-first century: A perspective," *Proc. IEEE*, vol. 85, pp. 1622–1638, October 1997.

[246] C. E. Shannon, "A mathematical theory of communication," *Bell Sys. Tech. J.*, vol. 27, pp. 379–423, 623–656, 1948.

[247] J. Shapira, "Microcell engineering in CDMA cellular networks," *IEEE Trans. Veh. Tech.*, vol. 43, pp. 817–825, November 1994.

[248] J. J. Shynk and R. P. Gooch, "Convergence properties of the multistage CMA adaptive beamformer," in *Proc. Asilomar Conf. Sign., Syst. Comp.*, pp. 622–626, 1993.

[249] B. Sklar, "Defining, designing and evaluating digital communication systems," *IEEE Commun. Mag.*, vol. 31, pp. 92–101, November 1993.

[250] B. Sklar, "Rayleigh fading channels in mobile digital communication systems Part I: Characterization," *IEEE Commun. Mag.*, pp. 90–100, July 1997.

[251] B. Sklar, "Rayleigh fading channels in mobile digital communication systems - Part II: Mitigation," *IEEE Commun. Mag.*, pp. 148–155, September 1997.

[252] E. Sousa, "Delay spread measurements for the digital cellular channel in Toronto," *IEEE Trans. Veh. Tech.*, vol. 43, pp. 837–847, November 1994.

[253] R. Steele, "Speech codecs for personal communications," *IEEE Commun. Mag.*, pp. 76–83, November 1993.

[254] R. Steele, J. Whitehead, and W. Wong, "System aspects of cellular radio," *IEEE Commun. Mag.*, pp. 80–86, January 1995.

[255] G. Stüber, *Fundamentals of Mobile Communications*. Kluwer Academic Publishers, 1998.

[256] B. Suard, A. Naguib, G. Xu, and A. Paulraj, "Performance of CDMA mobile communication system using antenna arrays," in *Proc. IEEE ICASSP*, pp. IV-153–IV-156, 1993.

[257] H. Sugimoto, L. K. Rasmussen, T. J. Lim, and T. Oyama, "Mapping functions for successive interference cancellation in CDMA," in *Proc. IEEE VTC*, pp. 2301–2305, May 1998. Ottawa, Canada.

[258] H. Suzuki, "A statistical model for urban multipath channels with random delay," *IEEE Trans. Commun.*, pp. 9–16, February 1972.

[259] T. Suzuki and Y. Takeuchi. "Real-time decorrelation scheme with multistage archietecture for asynchronous DS/CDMA," *Tech. Rep. IEICE, SST96-10, SAT96-24, RCS96-34, (in Japanese)*, June 1996.

[260] S. Swales, M. Beach, and D. Edwards, "Multi-beam adaptive base station antennas for cellular land mobile radio systems," in *Proc. IEEE VTC*, pp. 341–348, 1989.

[261] S. Swales, M. Beach, D. Edwards, and J. McGeehan, "The performance enhancement of multibeam adaptive basestation antennas for cellular land mobile radio systems," *IEEE Trans. Veh. Tech.*, vol. 39, pp. 56–67, 1990.

[262] J. Talvitie, *Wideband Radio Channel Measurement, Characterisation and Modeling for Wireless Local Loop Applications*. PhD thesis, University of Oulu, 1997.

[263] S. Talwar and A. Paulraj, "Performance analysis of blind digital signal copy algorithms," in *Proc. IEEE MILCOM*, pp. 123–127, 1994.

[264] S. Talwar, M. Viberg, and A. Paulraj, "Blind estimation of multiple cochnnel digital signals arriving at an antenna array," in *Proc. Asilomar Conf. on Sign., Syst. Comp.*, pp. 349–353, 1993.

[265] P. H. Tan, L. K. Rasmussen and T. J. Lim, "Constrained maximum-likelihood detection in CDMA," submitted to *IEEE Trans. Commun.*, June 1999.

[266] P. H. Tan and L. K. Rasmussen, "Interference cancellation in CDMA based on iterative techniques for solving linear systems," submitted to *IEEE Trans. Commun.*, April 1999.

[267] M. Tangeman, "Smart antenna technology for GSM/DCS1800," in *Proc. PWC*, 1996.

[268] V. Tarokh, A. Naguib, N. Seshadri, and A. R. Calderbank, "Low-rate multi-dimensional space-time codes for both slow and rapid fading channels," in *Proc. IEEE PIMRC*, (Helsinki, Finland), pp. 1206–1210, September 1997.

[269] V. Tarokh, N. Seshadri, and A. R. Calderbank, "Space-time codes for high data rate wireless communication: Performance criterion and code construction," *IEEE Trans. Inform. Theory*, vol. 44, pp. 744–765, March 1998.

[270] S. Tekinary and B. Jabbary, "Handover and channel assignment in mobile cellular networks," *IEEE Commun. Mag.*, vol. 29, pp. 42–46, November 1991.

[271] I. E. Telatar, "Capacity of multi-antenna Gaussian channels." submitted to *IEEE Trans. Inform. Theory*.

[272] J. Thompson, P. Grant, and B. Mulgrew, "Analysis of CDMA antenna array receivers with fading channels," in *Proc. IEEE ISSSTA*, (Mainz,Germany), pp. 297–301, University of Kaiserslautern, 1996.

[273] J. R. Treichler and B. G. Agee, "A new approach to multipath correction of constant modulus signals," *IEEE Trans. Acoust., Speech Sign. Proc.*, vol. ASSP-31, pp. 459–471, April 1983.

[274] J. R. Treichler and M. G. Larimore, "New processing techniques based on the constant modulus adaptive algorithm," *IEEE Trans. Acoust., Speech Sign. Proc.*, vol. 2, pp. 420–431, April 1985.

[275] M. Tsatsanis and G. Giannakis, "Blind estimation of direct sequence spread spectrum signals in multipath," *IEEE Trans. Signal Proc.*, vol. 45, no. 5, pp. 1241-1252, May 1997.

[276] G. Tsoulos, M. Beach, and J. McGeehan, "Wireless personal communications for the 21st century: European technological advances in adaptive antennas," *IEEE Commun. Mag.*, vol. 35, pp. 102–109, September 1997.

[277] K. Tsunekawa, "Diversity antenna for portable telephones," *IEEE Veh. Tech. Conf. '89, San Francisco, CA*, pp. 50–56, 1989.

[278] K. Tutschku and P. Tran-Gia, "Spatial traffic estimation and characterization for mobile communication network design," *IEEE Journ. Select. Areas Commun.*, vol. 16, pp. 804–811, June 1998.

[279] ETSI SMG2, "The ETSI UMTS terrestrial radio access (UTRA) ITU-R RTT candidate submission," *Tech. Rep. ETSI STC SMG2*, May/June 1998.

[280] R. L. Urbanke, *On Multiple-access Communication*. PhD thesis, Washington University, Sever Institute of Technology, 1995.

[281] ETSI SMG2, "UTRA physical layer description FDD parts," *Tech. Rep. v0.4, 1998-06-25, ETSI STC SMG2 UMTS-L1*, June 1998.

[282] M. C. Valenti, "Iterative detection and decoding for wireless communications." PhD thesis, Virginia Polytechnic, Blackburg, Virginia, July 1999.

[283] R. A. Valenzuela, "A ray tracing apporach for predicting indoor wireless transmission," in *Proc. IEEE VTC*, pp. 214–18, 1993.

[284] D. R. van Rheeden and S. C. Gupta, "A geometric model for fading correlation in multi-path radio channels," in *Proc. IEEE ICC*, (Atlanta, Georgia), pp. 1655–1659, June 1998.

[285] P. van Rooyen and R. Kohno, "DS-CDMA performance with maximum ratio combining and antenna arrays in Nakagami multipath fading," in *Proc. IEEE ISSSTA*, (Mainz, Germany), pp. 292–296, University of Kaiserslautern, 1996.

[286] D. J. van Wyk, I. J. Oppermann, and L. P. Linde, "Performance tradeoff among spreading, coding and multiple-antenna transmit diversity for capacity space-time coded DS/CDMA." in *Proc. IEEE MILCOM* (Atlantic City, U.S.A), 1999.

[287] D. J. van Wyk and L. P. Linde, "Design and performance evaluation of a Turbo/Walsh-Hadamard coded DS/CDMA system," *Trans. South African Inst. Elect. Eng.*, vol. 88, pp. 17–27, September 1998.

[288] D. J. van Wyk and L. P. Linde, "Turbo-coded/multi-antenna diversity combining scheme for DS/CDMA systems," in *Proc. IEEE ISSSTA*, (Sun City, South Africa), pp. 18–22, September 1998.

[289] D. J. van Wyk and L. P. Linde, "A turbo coded DS/CDMA system with embedded Walsh-Hadamard codewords: Coder design and performance evaluation," in *Proc. IEEE ISSSTA*, (Sun City, South Africa), pp. 359–363, September 1998.

[290] D. J. van Wyk and L. P. Linde, "Fading correlation and its effect on the capacity of space-time turbo coded DS/CDMA systems." in *Proc. IEEE MILCOM* (Atlantic City, U.S.A), 1999.

[291] D. J. van Wyk, L. P. Linde, and P. van Rooyen, "On the performance of a turbo-coded/multi-antenna transmission diversity scheme for DS/CDMA systems with adaptive channel estimation," in *Proc. IEEE ICT*, (Cheju, Korea), June 1999.

[292] D. J. van Wyk, I. J. Oppermann, and L. P. Linde, "Low rate coding considerations for space-time coded DS/CDMA," in *Proc. IEEE VTC*, (Amsterdam, The Netherlands), pp. 2520–2524, September 1999.

[293] M. K. Varanasi, "Parallel group detection for synchronous CDMA communication over frequency-selective fading channels," *IEEE Trans. Inform. Theory*, vol. 43, pp. 116–128, January 1996.

[294] M. K. Varanasi and B. Aazhang, "Multistage detection in asynchronous code–division multiple–access communications," *IEEE Trans. Commun.*, vol. 38, pp. 509–519, April 1990.

[295] R. G. Vaughan and J. B. Anderson, "Antenna diversity in mobile communications," *IEEE Trans. Veh. Tech.*, pp. 149-172, vol. VT-36, 1987

[296] R. Vaughan and N. Scott, "Closely spaced monopoles for mobile communications," *Radio Sci.*, vol. 28, pp. 1259–1266, 1993.

[297] S. Verdú, "Minimum probability of error for asynchronous Gaussian multiple access channels", *IEEE Trans. Info. Theory*, Vol. 32, pp. 85-96, January 1986.

[298] S. Verdú, "Optimum multi-user asymptotic efficiency", *IEEE Trans. Commun.*, Vol. 34, pp. 890-897, September 1986.

[299] S. Verdú, "The capacity region of the symbol-asynchronous Gaussian multiple access channel," *IEEE Trans. Inform. Theory*, vol. 35, pp. 733–751, July 1989.

[300] S. Verdú, "Recent progress in multiuser detection," *Proc. Int. Conf. Advanc. Commun. Contr. Sys.*, pp.66-77, October 1988.

[301] S. Verdú, "Computational complexity of optimum multiuser detection," *Algorithmica*, vol. 4, pp. 303–312, 1989.

[302] S. Verdú, *Multiuser Detection*. Cambridge University Press, 1998.

[303] A.J. Viterbi, "Error bounds for convolutional codes and an asymptotically optimum decoding algorithm", *IEEE Trans. Info. Theory*, Vol. IT-13, pp 260-269, April 1967.

[304] A. J. Viterbi, "Orthogonal tree codes for communication in the presence of white Gaussian noise," *IEEE Trans. Commun.*, vol. COM-15, pp. 238–242, April 1967.

[305] A. J. Viterbi, "Spread spectrum communication — myths and realities," *IEEE Commun. Mag.*, pp. 11–18, July 1979.

[306] A. J. Viterbi, "When not to spread spectrum — a sequel," *IEEE Commun. Mag.*, vol. 23, pp. 12–17, April 1985.

[307] A. J. Viterbi, "Very low rate convolutional codes for maximum theoretical performance of spread-spectrum multiple-access channels," *IEEE Journ. Select. Areas Commun.*, vol. 8, pp. 641–649, May 1990.

[308] A. J. Viterbi, *Principles of Spread Spectrum Communication*. Addison-Wesley Publishing Company, Massachusetts, 1995.

[309] A. J. Viterbi and J. K. Omura, *Principles of Digital Communication and Coding*. McGraw-Hill, 1979.

[310] B. Vucetic and Y. Sato, "Optimum soft-output detection for channels with intersymbol interference," *IEEE Trans. Inform. Theory*, vol. 41, pp. 704–713, May 1995.

[311] J.Walfisch and H. Bertoni, "A theoretical model of UHF propagation in urban environments," *IEEE Trans. Ant. Prop.*, vol. 36, pp. 1788–1796, December 1988.

[312] C. Ward, M. Smith, A. Jeffries, D. Adams, and J. Hudson, "Characterising the radio propagation channel for smart antenna systems," *Elect. & Commun. Eng. Journ.*, pp. 191–201, August 1996.

[313] W. Webb, "Modulation methods for PCN," *IEEE Commun. Mag.*, pp. 90–95, December 1992.

[314] W. Webb, "Sizing up the microcell for mobile radio communications," *Elect. & Commun. Eng. Journ.*, pp. 133–140, June 1993.

[315] V. Weerackody, "Diversity for direct-sequence spread spectrum system using multiple transmit antennas," in *Proc. IEEE ICC*, (Geneva, Switzerland), pp. 1775–1779, May 1993.

[316] L. Wei and L. K. Rasmussen, "Near ideal noise whitening filter for an asynchronous time-vaying CDMA system," *IEEE Trans. Commun.*, vol. 44, pp. 1355–1361, October 1996.

[317] S. Wijayasuriya, J. Norton and J. McGeehan, "A sliding window decorrelating receiver for multiuser DS-CDMA mobile radio networks," *IEEE Trans. Veh. Technol.*, vol. 45, pp. 503–521, August 1996.

[318] C. Wijffels, H. Misser, and R. Prasad, "A micro-cellular CDMA system over slow and fast Rician fading radio channels with forward error correction coding and diversity," *IEEE Trans. Veh. Tech.*, vol. 42, pp. 570–580, November 1993.

[319] N. Wilberg, H.-A. Loeliger, and R. Kötter, "Codes and iterative decoding on general graphs," *European Trans. Telecommun.*, vol. 6, pp. 513–525, September-October 1995.

[320] J. H. Winters, "Switched diversity with feedback for DPSK mobile radio systems," *IEEE Trans. Veh. Tech.*, vol. IT-32, pp. 134–150, February 1983.

[321] J. H. Winters, "Diversity gain of transmit diversity in wireless systems with Rayleigh fading," in *Proc. IEEE ICC*, (New Orleans, LA), pp. 1121–1125, May 1994.

[322] J. Winter, J. Salz, and R. Gitlin, "The impact of antenna diversity on the capacity of wireless communication systems," *IEEE Trans. Commun.*, vol. 42, pp. 1740–1750, February/March/April 1994.

[323] A. Wittneben, "A new bandwidth efficient transmit antenna modulation diversity scheme for linear digital modulation," in *Proc. IEEE ICC*, (Geneva, Switzerland), pp. 1630–1634, May 1993.

[324] B. D. Woerner and W. E. Stark, "Trellis-coded direct-sequence spread-spectrum communications," *IEEE Trans. Commun.*, vol. 42, pp. 3161–3170, December 1994.

[325] A. Wojnar, "Unknown bounds on performance in Nakagami channels," *IEEE Trans. Commun.*, vol. COM-34, pp. 22–24, January 1986.

[326] Z. Xie, R. T. Short, and C. K. Rushforth, "A family of suboptimum detector for coherent multiuser communications," *IEEE Journ. Select. Areas Commun.*, vol. 8, pp. 683–90, May 1990.

[327] S. Yasuda, "Effect of adaptive frequency re-use in mobile communication dynamic channel assignment," *Proc. IEICE Spring Nat. Conv.*, vol. B.860, p. 860, 1989.

[328] M. Yokoyama, "Decentralisation and distribution in network control of mobile radio communications," *IEICE Trans.*, vol. E73, pp. 1579–1586, October 1990.

[329] O. Yue, "Design trade-offs in cellular/PCS systems," *IEEE Commun. Mag.*, vol. 34, pp. 146–152, September 1996.

[330] P. Zetterberg and B. Ottersten, "The spectrum efficiency of a base station antenna array system for spatially selective transmission," in *Proc. IEEE VTC*, 1994.

[331] "Mobile Communication with Base Station Antenna Array: Propagation Modelling and System Capacity." Tech. Report, Royal Institute of Technology, January 1995.

[332] P. Zetterberg and P. L. Espensen, "A downlink beam steering technique for GSM/DCS1800/PCS1900," in *Proc. IEEE PIMRC*, October 1994.

[333] P. Zetterberg, P. L. Espensen and P. Mogensen, "Propagation, beam steering and uplink combining algorithms for cellular systems ." in *Proc. ACTS Mobile Commun. Summit*, Garanda, Spain, November, 1996.

[334] E. Zehavi and A. J. Viterbi, "On new classes of orthogonal convolutional codes," *IEEE Commun., Contr. Sig. Proc.*, pp. 257–263, 1990.

[335] J. Zou and V. Bhargava, "Design issues in a CDMA cellular system with heterogeneous traffic types," *IEEE Trans. Veh. Tech.*, vol. 47, pp. 871–884, August 1998.

[336] T. Zwick, D. Didascalou, and W. Wiesbeck, "A broadband statistical channel model for SDMA applications," in *Proc. IEEE ISSSTA*, (Sun City, South Africa), pp. 527–531, University of Pretoria, September 1998.

Index

ACTS, 11
Adaptive beamforming algorithms, vi, 96, 98
Adaptive MMSE detection, vii, 165
ADC, 40
ADSL, 1
Advantages of space-time processing, v, 35
AMPS, 18
Angular dispersion, 55, 63
Angular distribution, 43, 58, 60, 70
Angular spread, xvii, 36, 56, 58, 72–74, 87–88
  rule of thumb, 73
Angular subscriber distribution, vi, 59
Antenna element
  optimum spacing, 129
Antenna pattern, 19, 34–35, 44, 56, 58–59, 107, 117, 126, 23
Antenna sectorization, 17
Antenna selection TDTD, 115, 206
Antenna selection, 116
Antenna weight, 97, 23, 97, 112
APON, 12
Array factor, 27
Array geometry, 25
AS-TDTD, 115–116, 206, 213
Asymmetric bandwidth, 8
ATM, 1
Average direct power, 119
Average scattered power, 119
B-ISDN, 12
B-JOR, 172
Backward recursion, 252
Bad urban, 69, 86
Bandwidth
  efficiency, xiv, 150
  requirements, 196
Beam pattern
  correlation influence, 107
Beam steering, 44, 82

Beamforming, 30, 133, vi, xi, 13–14, 17, 23, 27–28, 30, 33–34, 36, 40, 93, 96
  algorithms, 98
  blind, 99
  CDMA, 99, 103
  discrete signal structure, 102
  DOA estimation, 101
  general, 102
  non-blind, 98
  property-restoral techniques, 101
  fixed, 96
  one-shot, 97
  performance, 125, 140
Bessel functions, 68
Blind beamforming, vii, 98–100
Blind transmit diversity, 112
Block iterative method, 170–171
Block-wise JOR, 172
Bore-sight direction, 24
Branch transitions probabilities, 252
Broadband satellite, v, 11
Broadband service, 12, xvii
Broadside arrays, 24–25
BSS, 10
CDSM fading correlation, xii, 69, 68
CDSM, vi, 64–65, 68–70, 73–74, 87, 90
CDTD, vii, 113, 116, 184, 192–194, 198
  block diagram, 113
  delayed, 114
Cellular channel model, xii, 89
Cellular spectral efficiency, 17–20
Channel model
  LOS
    correlation, 107
  macro-cell
    correlation, 111
  NLOS
    correlation, 107
Channel models, vi, xii, 44, 46, 55, 57–58, 63–65, 67, 69, 71–73, 75, 77, 79, 81,

83, 85, 87, 89, 91, 125
Channel state (side) information, 184
Circular array, xii, xvii, 109
Circular disk of scatterers model, vi, 65
  geometry, xii, 66
CL-AS, 214
Clipped soft decision, 169
Closed-loop antenna selection, 214
Clustered reflectors, 64
CMA, vii, 99
Co-channel interference, 19–20, 35, 96, 118
Code-division transmit diversity, vii, 113
Coded time-division transmit diversity, xiii, 116
Coded transmit diversity, xv, 185, 206
Coding gain, 186
Coding, 13, 15, 17, 20, 30, 34, 54, 99, 113
Code-diversity gain, 184
Coherence time, 51–53, 116
Computational complexity, 26, 102, 104, 154, 166–167, 174, 189, 191
Concatenated convolutional codes, 187
Constant correlation model, 237, 240
Constant modulus algorithm, vii, 99
Constraint length, 161
Conventional detection, vii, 156
Convolutional coding, 161, 184, 192, 211, 267
  bounds, 196
  CDTD, 211
  concatenated, 187
  low rate, 186, 204
  orthogonal, 184
  rate-compatible, 186
  recursive systematic, 188
  super-orthogonal, 184
  TDTD, 211
Correlation, 28–29, 38, 44, 46, 49–50, 52–53, 63–65, 68, 72, 74, 76–77, 79, 81, 87, 90, 94, 97–98, 102–103, 107, 117–118, 128, 130, 132, 137–140, 142, 144–145, 154, 164, 167–168, 181–182, 199, 210, 237
  ULA beam pattern, 107
Correlator receiver, 30, 154, 194
Cost function, 96–100, 102, 165
CSD, 169
CSI, 184, 199
CT-2, 18
CTM, 12
DCS 1800, 10, 5
DDFD, 167
Decorrelating detector, vii, 162
Decorrelating noise whitening filter, 167
Decorrelating receiver, 30
Delay diversity, 112, 183
Delay spread, 21, 48–49, 55, 63, 79, 85

Delayed CDTD, 114
Diffraction, 45, 85–86
Discrete memoryless source, 191
Discrete uniform distribution, vi, 76
Distant scatterers, 64
Distributed antennas, 24
Diversity antenna arrays, 179
Diversity branches, 117, 120
Diversity combiner, 117
Diversity gain, 117, 123
Diversity performance, 120
Diversity
  antenna-selection, 183
  basic structure, 118
  frequency-offset, 183
  performance, 140, 120
  phase sweeping, 183
DOA distribution, 72
DOA estimation, vii, 101
Doppler
  power spectrum, vi, 50, 53, 51, 38, 51–52
  shift, 51, 55, 78, 82
  spread, 51–52, 63
Downlink CDTD, 113
Downtown clustered micro-cells, 22
DQPSK, 65
DSP, 40
DUD model geometry, xii, 77
DUD, vi, 76
EFD, vi, 87
Effective scatterer model, vi, 75
Electrical field, 27
End-fire arrays, 24–25
Envelope correlation
  rural macro-cell, 72
  urban micro-cell, 72
Equal gain combining, 117, 157
ERC, 4
Erlang, 18, 122
ESM geometry, xii, 76
ESM, vi, 75
ESPRIT, 35, 101
Estimation algorithm, 58
ETSI, 9, 5
Euclidean distance, 174
Excitation amplitude, 25
Excitation phase, 25
Exponential correlation model, 237
Exponential fading distribution, 87
Extended tap-delay-line model, vi, 84
Extrinsic information, ix, 248–249
Fading distribution, vi, 86
Fading parameter, 138
Fast fading, vi, 55, 94, 113
FDMA, 18–19
Fixed beamforming, vi, 96–97
Flat fading, vi, 50, 54, 114

Forward recursion, 252
FPLMTS, 2
Free space loss, 44, 84
Frequency-selective fading, vi, 54
FTF, 41
FTTB, 12
FTTC, 12
FTTH, 12
GAA, 65, 81
Gamma distribution
　correlated multivariate, 237
Gauss-seidel inner iteration, vii, 172
Gaussian bell shape, 65, 69, 72
Gaussian fading distribution, 87
Gaussian scatterer model, 14, 69
　correlation, 74
Gaussian wide sense stationary uncorrelated
　model, 79
GBCM, vi, 78
GBSB, vi, 77
GBSBEM, vi, xi, 47–48, 50, 52–53, 78
Geometrical theory of diffraction, 85
Geometrically based circular macro-cell
　model, 78
Geometrically based elliptical micro-cell
　wideband model, 78
Geometrically based single-bounce
　statistical models, 77
GFD, vi, 87
GS, vi, xii, 69–70, 107
GSM, 10, 5–6, 15
GWSSUS geometry, xii, 80
GWSSUS, vi, 79
Hamming weight, 188
Hand overs, 20
HFC, 12
HFR, 12
High-rank channel, vi, xii, 56–57, 55–56, 58,
　88, 90
Hilly terrain, 70
Hot spots, 22
HSR, 19–20, 33
Hyperbolic tangent, 169–170, 173
IC, 154
IMT-2000, 2–v, 7, 9, 12, 17
IN, 10
Indoor environments, 18, 70, 8–9
Information society, 7
Intelligent technology, 34
Interference cancellation, vii, 167, 154
　linear, 170
　non-linear, 172
Interleaver gain, 189
Interleaver
　block interleaving, 189
　circular shift, 189
　semi-random, 189

Internet, 10, 1–2, 11, 183
Intrinsic information, ix, 248
IP, 8
IPI, 54
IS-54, 5, 18
IS-95, 5, 18
ISDN, 10, 1, 12
ISI, 112
Isotropic antenna, 24
Iterative decoding algorithm, ix, 250
Iterative methods
　Gauss-Seidel, 163
　Jacobi, 163
Iterative multiuser detectors, 154
Iterative turbo decoding, 247
ITU, 2–3
Jacobi iteration, vii, 171–172
Jacobi over-relaxation, 172
Joint decoding, viii, 173
Joint multiuser detection, 154
JOR, 172
LAN, 8
Layered space-time coding, 187, 184, 193,
　214
　low rate code extensions, 195
Layered space-time transmit diversity, 220
Least squares, 100
LEO, 12
Linear detection, vii, 162
　adaptive MMSE, 165
　decorrelating detector, 162
　MMSE detector, 163
Linear interference cancellation, vii, 170
　Gauss-Seidel Inner Iteration, 172
　Jacobi inner iteration, 171
Linear MMSE detector, 181, 163
LLR, 248
LMDS, 12
LMMSE detector, 164–165
LMMSE, 163–164
LMS, 14, 41, 98, 166–167
LMS-MMSE detector, 181
Local scattering elements, 125
Local scattering environment, vi, 64, 69, 87
Log-likelihood metric, 162
Log-likelihood ratio, 248
Log-MAP, 188
Log-normal shadowing, 45, 268, 271
LOS, 45–46, 48, 50–53, 56, 58, 64, 71–73, 76,
　86–88, 107, 109, 133, 139, 142, 145
　performance, 140
Low rate coding, 184, 186
Low-rank channel, vi, xii, 56, 58, 55, 58, 88
LS, 100
LS-DRMTCMA ULA beampattern, xii,
　105–106
LSL, 41

Macro-cell, 21, 58, 68, 72, 68, 72–73, 107
Mahalanobis distance, 174
Main lobe, 58
MAP detection, 175, 162, 188, 191, 247
Max-log-MAP, 188
Maximum ratio combiner, 119, 94
Maximum excess delay, 79, 46, 48–49, 54, 79
Maximum free distance, 186
Maximum *a posteriori*, 188, 247
Maximum-likelihood sequence detector, 154
Measurement-based channel models, vi, 84
Metcalfe's Law, 1
MFD, 186
Micro-cell, 21, 47, 68, 72, xii, 61, 65, 69–73, 78
MIMO, 28–29
Minimum mean square error, 97, 163
MIP, 46–47, 51, 69, 71, 128, 136
MLSD, 154, 174
MM LEO, 12
MMDS, 12
MMSE detector, 166, vii, 155, 163, 165, 181
MMSE receiver, 164
MMSE solution, 164
MMSE, 30, 97–98, 163
Model
 fading correlation, 74
Modeling
 scattering elements, 70
Modified Saleh-Valenzuel model, vi, 83
Modulation
 multiuser, 147
Moore's law, 1
MoU, 5
MRC, 94, 117, 157
MU-MISO, 29
MU-SIMO, 30
MU-SISO, 29
MUD, 13
Multi-band, 5, 40
Multi-mode, 5
Multi-path intensity profile, v, 46
Multi-stage detector, 168
Multi-stage non-linear SIC, 168
Multimedia, 10, 5, 8–9, 11
Multiple access transmit diversity, 193
Multiuser detection, vii, 97, 153–157, 159, 161, 163, 165, 167, 169, 171, 173, 175, 177, 179, 181
 adaptive MMSE, 165
 clipped soft decision, 169
 decision statistic, 155
 decorrelating decision feedback detector, 167
 hard decisions, 169
 interference cancellation, 167
 linear detection, 162

linear interference cancellation, 170
mapping functions, 169
MLSD, 154
MMSE detector, 163
non-linear interference cancellation, 172
one-shot detection, 156
optimal detection, 160
practical implementation, 155, 168, 175, 181
soft decisions, 169
speed of convergence, 166
system model, 155
Multiuser detector, 153–154
 interference cancellation, 154
 one-shot, 154
MUSIC, 35, 101
Nakagami fading, 118–119, 46
Nakagami-m, 37, 46, 87
NLOS, 67–68, 72–74, 107
NO-CDTD, 113–115, 213
Non-blind beamforming, vii, 98
Non-linear interference cancellation, vii, 172
Non-linear SIC, 168
Non-orthogonal code-division transmit diversity, 113
Null steering, 23, 153
O-CDTD, 113–114, 206, 211, 213
Omni-directional antennas, 63, 24
One-shot beamforming, 97, vi, 97
One-shot detection, 156
 conventional detection, 156
 decorrelating decision feedback detector, 167
 linear detection, 162
 optimal detection, 160
One-shot multiuser detectors, 154
Optimal detection, vii, 160, 162
Optimal linear ML detector, 154
Optimal ML detector, 154
Optimal successive decoder, 194
Orthogonal CDTD, 206
Orthogonal code-division transmit diversity, 113
Orthogonal convolutional codes, 184, 186, 195
Orthogonal convolutional encoder, 196
Orthogonal spreading codes, 158
OSD, 194
Outage probability, 36, 118
Outdoor, 9, 213
Pairwise bit error, 201
Pairwise error probability, 197–201
Path diversity, 116
Path loss, 42–45, 71, 78, 84–85
Pay-per-bit, 8
PCS, 10, 5, 13
PDNW, 167

Phased arrays, 25
Pico cell, 90
Pilot symbol assisted, 192
Planar arrays, 24–25
Polynomial time, 161, 182
Power control, 33, 41–42, 116, 129, 175, 183, 193, 258, 265
Power delay profiles, viii, 63, 70, 233, 235
Pre-RAKE combining, xiii, 117
　CDTD, 116
　TDTD, 116
Pre-RAKE, 116
　block diagram, 116
Processing gain, 33, 37, 148, 154, 160, 175, 206, 264
Propagation path, v, 14, 43, 46, 50, 64, 111
PSA, 192
QoS, 9
RACE, 11
Radio channel units, 120
Radius of scatterers, 94
RAKE receivers, 30, 139
Raleigh's model, vi, xii, 81–82
Rate of convergence, 26
Rate-compatible convolutional codes, 186
Ray tracing models, vi, 85
Rayleigh fading, 119, 199
Rayleigh, 37, 46, 65, 81–83, 86, 90, 94, 120, 184, 202, 204, 206
RCCC, 186
RCU, 120, 122
Re-configurable terminals, v, 9
Re-use cluster, 18, 123
Receive diversity, vii, 93–94, 107, 111, 113, 117, 137, 139, 157, 183, 192
　performance, 137
Recursive systematic convolutional encoders, 188
Reflection, 45, 70, 72, 85–86
Repetition coding, 267
Rician, 46, 86, 90, 119
RLS, 167
Round-robin antenna selection, 112
Round-robin TDTD, 115
RR-TDTD, 115–116, 213
RSC, 188
Rural, 9, 46, 64–65, 69, 72, 84
S-UMTS, 12
Scatterers, 14, 36, 44, 47, 50, 56, 58, 64–65, 67, 71, 73–80, 82
Scattering element
　standard deviation
　　macro-cell, 72
Scattering environment, vi, 14, 43–44, 47, 52–53, 55–56, 58, 64, 73, 86–88, 90, 93–94, 96, 107, 115, 138, 184
Scattering points, 65, 69, 85

Scattering, 14, 35, 44–47, 56, 58–59, 63–65, 67–75, 90
SDH, 12
SDMA, xi–12, 17–18, 20–21, 33, 36, 94, 107
Second generation, 2–3, v–6, 8–9, xi
Sectorization, vii, xiii, 120–122
Selection diversity combiner, 118
Selection diversity, 157, 214
Service download, v, 11
SFIR, 19–20, 33, 59
Shadowing, 42–45
Shannon bound, 187
Signal processing techniques, 17
SIM, 10–11
SINR, 33–34, 36, 41, 122, 164
SISO, xvi, 188, 220, 248–250
Slow fading, vi, 55
Smart antenna systems, v, 40
Smart antenna techniques, vi, 93, 95, 97, 99, 101, 103, 105, 107, 109, 111, 113, 115, 117, 119, 121, 123
Smart antenna, xii, 95
　performance, 125
Smart antennas, 17
SNR, 19, 54–55, 99, 103, 117, 120, 188–189, 198, 202, 204, 220
SOCC, 184, 186, 195
Soft failure, 195
Soft output Viterbi algorithm, 188
Software radio, 40–41
SOTC, 184, 186, 196
SOTTD block diagram, xvi, 219
SOTTD, viii, 214, 218
SOVA, 188, 191
Space diversity, 116, 184
Space-frequency correlation, vi, 49
Space-frequency transmit diversity, 112
Space-time base station, v, 34
Space-time channel models, vi, 87
Space-time coded transmit diversity, 220
Space-time codes, 15
Space-time coding, 14, 115, 125, 183, 187
　analytic, 202
　convolutional, 193
　　block diagram, 194
　correlation effects, 206
　extensions, 214
　layered, 193
　　block diagram, 193
　multiple access, 184
　performance, 196
　simulation, 211
　system model, 191
　transmit diversity, 183, 191
　turbo, 193
Space-time correlation, vi, 52
Space-time diversity, 206

Space-time gain, 184
Space-time mobiles, v, 35
Space-time processing, 17, v, 12, 14, 17, 28, 126
Space-time techniques, v, 11
Space-time transmit diversity, 112, 184, 196, 220
Space-time turbo diversity, 184
Space-time, vi, 11, 13–14, 17–20, 23–24, 28, 30, 33–36, 43–44, 46–49, 55, 58–59, 63–65, 67, 69, 71, 73–75, 77, 79, 81, 83, 85, 87, 89, 91, 96, 99, 107, 120, 125, 151, 183–184, 187, 191–193, 196, 200–201, 211
Spatial filtering, 19, 126, 19
Spatial RAKE combiner, 154
Spatial signal processing, 17
Spatial-temporal, 86
Spectral density, 38, 46, 51
Standardization, 2–3
Statistically optimal, 97
Steepest decent algorithm, 166
Steering vector, 27, 35, 80, 94, 102–103, 128
SU-MISO, 29
SU-SIMO, 30
SU-SISO, 29
Suburban, 61, 65, 84, 86
Super-orthogonal convolutional codes, 184, 195
Super-orthogonal turbo codes, 184
Super-orthogonal turbo transmit diversity, viii, 214, 218
Switched beam, 13
Switched data, 5
Switched diversity, 112
Switched-beam, xiii, 121, vii, 120–122
Tail termination, ix, 254
TDMA, 18–19, 23, 111
TDTD, vii, xiii, 113, 115–116, 184, 192–193, 198
  antenna-selection, 115
  block diagram, 116
  round-robin, 115
Temporal fading model, xvii, 88
Temporal fading, v, 45, 136
Temporal signal processing, 17
Temporal, 17, 43, 55, 86, 90, 99, 133, 192, 206, 220
Terrestrial radio access, 9
Terrestrial, 2, 4–6, 8, xi, 15, xvii
Third generation, v, 7, 9, 12, 14, 17
Time-division transmit diversity, vii, 113, 115
Time-only receiver, 30
Time-variant, 46, 50–51, 53, 85
TOA, xi, 47–50, 52–53, 77–78
Total path loss, 45

TPC, 265
Tracking, 20, 26, 35, 198
Transmit diversity order, 206
Transmit diversity, vii, xv, 93, 96, 111–113, 183–184, 186, 192–193, 201, 206, 211–212, 265, 267, 272
  blind, 112
  code division, 113
  for CDMA, 113
  space-phase, 112
  space-space, 112
  time division, 115
  with feedback, 112
  with hybrid feedback/training, 112
  with training information, 112
Trellis coding, 186
TSUNAMI, 11, 69
TTD block diagram, xvi, 215
TTD, viii, 214
Turbo code input-output CPDF, viii, 243, 245
Turbo code, 187
  conditional pdf, 244
  constituent encoder, 191
  decoding, 191
  extrinsic information, 191
  interleaver, 189
  iterations, 191
  performance, 189
  permuter, 189
  puncturing, 191
  weight enumeration recursion, 243
Turbo coded transmit diversity, 206
Turbo codes, 184, 186–188
Turbo coding, viii, 13, 174, 186–187, 192, 211, 267
  backward recursion, 252
  bounds, 201
  branch metric, 251
  CDTD, 211
  decoding, 247
    MAP, 247, 250
  extrinsic information, 248
  forward recursion, 252
  intrinsic information, 248
  iterative decoding, 247, 250
  LLR, 253
  low rate, 204
  rate-$1/(Z+1)$, 214
  SISO decoding, 248
  tail termination, 254
Turbo decoding, viii, 174, 188
Turbo encoding, viii, 188
Turbo transmit diversity, viii, 214
Typical urban, xii, 61, 69, 86
ULA, xi, 28–29, 96, 104, 107, 129–130, 132, 136, 145

UMTS vehicular channel model, xiv–xv, 180
UMTS, 10, 4–9, xi–12, 15, xvii, 22, 35, 55,
    151, 154, 175, 177, 179, 260
Uniform arrays, 24–25
Uniform distance distribution, 61
Uniform linear array, 25
Uniform sectored distribution model, 82
Unpaired bands, 4
USD geometry, xii, 83
USD, vi, 82
UTRA, 4, 9
V-SAT, 12
VHE, 10, 8
Virtual connectivity, 8
Viterbi algorithm, 154, 161–162, 247,
    250–251
VPL, 86
W-ATM, 12
W-LAN, 12
Walsh-Hadamard orthogonal modulator, 218
Walsh-Hadamard matrix, 196
WCDMA Simulation Environment, ix, 257,
    263, 265, 267, 269, 271
WCDMA, 7, 55, 211, 220, 257–258
Weight vector, 39, 80–81, 41, 96, 99, 103
WLL, 12
XDSL, 12Ω

About the Authors

Pieter van Rooyen received the B.Eng, M. Eng (*cum laude*) and D.Eng degrees from the Rand Afrikaans University, Johannesburg, South Africa. Following two years in the research labs of Alcatel Altech Telecoms, he joined the University of Pretoria where he is currently a Professor in the Department of Electrical and Electronic Engineering. He is Director of the Alcatel Research Unit for Wireless Access (ARUWA) at the University of Pretoria. The aim of ARUWA is to conducts research into various aspects of mobile communication systems, including space-time processing. During 1995 he has been a Japanese Society for the Promotion of Science post-doctoral fellow at the Yokohama National University. Dr. van Rooyen has been chairman of the Fifth IEEE International Symposium on Spread Spectrum Techniques and Applications (ISSSTA'98) and has served as session chairman at various international conferences. He has published a number of papers in the areas of digital communications and smart antennas and is the co-author of patents in digital communications. Dr. van Rooyen co-edited a book entitled *"CDMA Techniques for 3rd Generation Mobile Systems"* published by Kluwer in 1998.

Michiel P. Lötter was born in 1971 in Pretoria, South Africa. He received the B.Eng and M.Eng degrees, both with honors, in electronic engineering from the University of Pretoria, Pretoria, South Africa, in 1992 and 1995, respectively. During 1997-1999, he completed a Ph.D. at the same university in the area of CDMA and Smart Antenna techniques. From 1995 to mid 1999, he was a development engineer with the Engineering Division of Alcatel Altech Telecoms, South Africa, where he conducted research into spread spectrum and CDMA radio systems. Also, during the period 1992-1994 he was a Research Assistant with the Laboratory for Advanced Engineering, Pretoria, South Africa. Currently, Dr. Lötter is with Altech in South Africa. Dr. Lötter was a Vice-Chairmen of ISSSTA'98 which was held in Sun City, South Africa. He has been a Guest Editor for a Special Issue of the Transactions of the South African Institute of Electrical Engineers on CDMA Technology, and has edited a book titled *"CDMA Techniques for 3rd Generation Mobile Systems"* with F. Swarts, P. van Rooyen and I. Oppermann. Dr. Lötter is the author and co-author of four patents in digital communications, an has published a number of articles in international journals. His research interests include CDMA and smart antennas for mobile and satellite communications. Dr. Lötter is a member of the IEEE and a registered professional engineer in South Africa.

Danie van Wyk was born in Bloemfontein, South Africa on May 18, 1971. In 1993 he received the B.Engineering (Electronic) degree (*cum laude*) from the University of Pretoria. In 1995 and 1996 he obtained the his honours and M.Sc degrees, respectively, both with distinction at the same university. Mr. Van Wyk is currently employed by Defencetek at the CSIR in Pretoria, where he works as systems engineer on electronic warfare, communications and navigation related military projects. He is currently working towards his Ph.D. in the area of space-time coding for mobile communications. His current interests include power and bandwidth efficient modulation and coding techniques for mobile digital communication, detection, navigation, synchronization, equalization, and channel modeling and simulation. He has published over 20 papers in the area of digital communications with emphasis on modulation and coding. Mr. Van Wyk is a member of the SAIEE and IEEE.

DISK DISCLAIMER

Copyright 1999, Kluwer Academic Publishers. All Rights Reserved.

This disk is distributed by Kluwer Academic Publishers with *ABSOLUTELY NO SUPPORT* and *NO WARRANTY* from Kluwer Academic Publishers.

Use or reproduction of the information provided on this disk for commercial gain is strictly prohibited. Explicit permission is given for the reproduction and use of this information in an instructional setting provided proper reference is given to the original source.

Kluwer Academic Publishers shall not be liable for damage in connection with, or arising out of, the furnishing, performance or use of this disk.